Barron's Regents Exams and Answers

Algebra I

Gary M. Rubinstein

B.A. Mathematics
Tufts University
M.S. Computer Science
University of Colorado

BARRON'S

Barron's Educational Series, Inc.

Published by Kaplan, Inc., d/b/a Barron's Educational Series
750 Third Avenue
New York, NY 10017
www.barronseduc.com

ISBN: 978-1-4380-0665-9
ISSN: 2376-7219

Printed in Canada

9 8 7 6 5 4 3 2 1

Kaplan, Inc., d/b/a Barron's Educational Series print books are available at
special quantity dipscounts to use for sales promotions, employee premiums,
or educational purposes. For more information or to purchase books, please
call the Simon & Schuster special sales department at 866-506-1949.

Contents

Regents Examinations, Answers, and Self-Analysis Charts   179

Preface

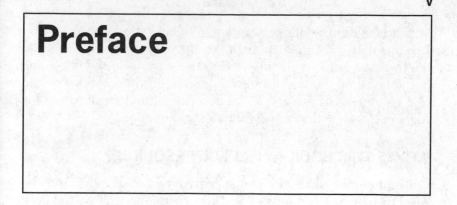

In June of 2014, New York State adopted a more difficult Algebra I course based on the Common Core curriculum and changed the Regents exam to reflect this new course. On the new Algebra I Regents exam, students are expected not only to solve algebra problems and answer algebra-related questions but also to *explain their reasoning*. This is a major shift in the testing of algebra concepts, so a thorough review is even more critical for success.

This book will introduce you to the concepts covered throughout the Algebra I Regents course as well as to the different types of problems you will encounter on the Regents exam at the end of the year.

WHO SHOULD USE THIS BOOK?

Students and teachers alike can use this book as a resource for preparing for the Algebra Regents exam.

Students will find this book to be a great study tool because it contains a review of Algebra I concepts, useful examples, and practice problems of varying difficulty that can be practiced throughout the school year to reinforce what they are learning in class. The most ideal way to prepare for the Algebra Regents exam is to work through the practice problems in the review sections and then the recently administered Regents exams at the end of the book.

Teachers can use this book as a tool to help structure an Algebra I course that will culminate with the Regents exam. The topics in the book are arranged by priority, so the sections in the beginning of the book are the ones from which more of the questions on the test are drawn. There are 13 sections, dedicated to all topics for Algebra I, each with practice exercises and solutions.

WHY IS THIS BOOK A HELPFUL RESOURCE?

Becoming familiar with the specific types of questions on the Algebra I Regents exam is crucial to performing well on this test. There are questions in which the math may be fairly easy but the way in which the question is asked makes the question seem much more difficult. For example, the question "Find all zeros of the function $f(x) = 2x + 6$" is a fancy way of asking the much simpler sounding "Solve for x if $2x + 6 = 0$." *Knowing exactly what the questions are asking is a big part of success on this test.*

Algebra has been around for thousands of years, and its basic concepts have never changed. So fundamentally, this algebra curriculum is not very different from the algebra taught in schools two years ago, ten years ago, or twenty years ago. But the exam that follows this Common Core-based course, with more complicated ways of asking questions and presenting problems, requires a specifically presented study plan that's more important than ever.

Gary Rubinstein

How to Use This Book

This book is designed to help you get the most out of your review for the new Regents exam in Algebra I. Use this book to improve your understanding of the Algebra I topics and improve your grade.

TEST-TAKING TIPS

The first section in this book contains test-taking tips and strategies to help prepare you for the Algebra I Regents exam. This information is valuable, so be sure to read it carefully and refer to it during your study time. Remember: no single problem-solving strategy works for all problems—you should have a toolbox of strategies to pick from as you're facing unfamiliar or difficult problems on the test.

PRACTICE WITH KEY ALGEBRA I FACTS AND SKILLS

The second section in this book provides you with key Algebra I facts, useful skills, and practice problems with solutions. It provides you with a quick and easy way to refresh the skills you learned in class.

REGENTS EXAMS AND ANSWERS

The final section of the book contains actual Algebra I Regents exams that were administered in August 2015, June 2016, August 2016, June 2017, August 2017, and June 2018. These exams and thorough answer explanations are probably the most useful tool for your review, as they let you know what's most important. By answering the questions on these exams, you will be able to identify your strengths and weaknesses and then concentrate on the areas in which you may need more study.

Remember, the answer explanations in this book are more than just simple solutions to the problems—they contain facts and explanations that are crucial to success in the Algebra I course and on the Regents exam. Careful review of these answers will increase your chances of doing well.

SELF-ANALYSIS CHARTS

Each of the Algebra I Regents exams ends with a Self-Analysis Chart. This chart will further help you identify weaknesses and direct your study efforts where needed. In addition, the chart classifies the questions on each exam into an organized set of topic groups. This feature will also help you locate other questions on the same topic in the other Algebra I exams.

IMPORTANT TERMS TO KNOW

The terms that are listed in the glossary are the ones that have appeared most frequently on past Integrated Algebra and the most recently Algebra I exams. All terms and their definitions are conveniently organized for a quick reference.

Test-Taking Tips and Strategies

Knowing the material is only part of the battle in acing the new Algebra I Regents exam. Things like improper management of time, careless errors, and struggling with the calculator can cost valuable points. This section contains some test-taking strategies to help you perform your best on test day.

TIP 1
Time Management

SUGGESTIONS

- *Don't rush.* The Algebra I Regents exam is three hours long. While you are officially allowed to leave after 90 minutes, you really should stay until the end of the exam. Just as it wouldn't be wise to come to the test an hour late, it is almost as bad to leave a test an hour early.
- *Do the test twice.* The best way to protect against careless errors is to do the entire test twice and compare the answers you got the first time to the answers you got the second time. For any answers that don't agree, do a "tie breaker" third attempt. Redoing the test and comparing answers is much more effec-

tive than simply looking over your work. Students tend to miss careless errors when looking over their work. By redoing the questions, you are less likely to make the same mistake.

- *Bring a watch.* What will happen if the clock is broken? Without knowing how much time is left, you might rush and make careless errors. Yes, the proctor will probably write the time elapsed on the board and update it every so often, but its better safe than sorry.

The TI-84 graphing calculator has a built-in clock. Press the [MODE] to see it. If the time is not right, go to SET CLOCK and set it correctly. The TI-Nspire does not have a built-in clock.

TIP 2

Know How to Get Partial Credit

SUGGESTIONS

- *Know the structure of the exam.* The Algebra Regents exam has 37 questions. The first 24 of those questions are multiple-choice worth two points each. There is no partial credit if you make a mistake on one of those questions. Even the smallest careless error, like missing a negative sign, will result in no credit for that question. Parts Two, Three, and Four are free-response questions with no multiple-choice. Besides giving a numerical answer, you may be asked to explain your reasoning. Part Two has eight free-response questions worth two points each. The smallest careless error will cause you to lose one point, which is half the value of the question. Part Three has four free-response questions worth four points each. These questions generally have multiple parts. Part Four has one free-response question worth six points. This question will have multiple parts.

- *Explain your reasoning.* When a free-response question asks to "Justify your answer," "Explain your answer," or "Explain how you determined your answer," the grader is expecting a few clearly written sentences. For these, you don't want to write too little since the grader needs to see that you understand why you did the different steps you did to solve the equation. You also don't want to write too much because if anything you write is not accurate, points can be deducted.

Here is an example followed by two solutions. The first would not get full credit, but the second would.

Example 1

Use algebra to solve for x in the equation $\frac{2}{3}x + 1 = 11$. Justify your steps.

Solution 1 (part credit):

$\begin{aligned} \frac{2}{3}x + 1 &= 11 \\ -1 &= -1 \\ \frac{2}{3}x &= 10 \\ x &= 15 \end{aligned}$	I used algebra to get the x by itself. The answer was $x = 15$.

Solution 2 (full credit):

$\dfrac{2}{3}x+1=11$ $-1=-1$ $\dfrac{2}{3}x=10$ $\dfrac{3}{2}\cdot\dfrac{2}{3}x=\dfrac{3}{2}\cdot 10$ $1x=15$ $x=15$	I used the subtraction property of equality to eliminate the $+1$ from the left-hand side. Then to make it so the x had a 1 in front of it, I used the multiplication property of equality and multiplied both sides of the equation by the reciprocal of $\dfrac{2}{3}$, which is $\dfrac{3}{2}$. Then since $1 \cdot x = x$, the left-hand side of the equation just became x and the right-hand side became 15.

- *Computational errors vs. conceptual errors*

In the Part Three and Part Four questions, the graders are instructed to take off one point for a "computational error" but half credit for a "conceptual error." This is the difference between these two types of errors.

If a four-point question was $x - 1 = 2$ and a student did it like this,

$$x - 1 = 2$$
$$+1 = +1$$
$$x = 4$$

the student would lose one point out of 4 because there was one computational error since $2 + 1 = 3$ and not 4.

Had the student done it like this,

$$x - 1 = 2$$
$$-1 = -1$$
$$x = 1$$

the student would lose half credit, or 2 points, since this error was conceptual. The student thought that to eliminate the -1, he should subtract 1 from both sides of the equation.

Either error might just be careless, but the conceptual error is the one that gets the harsher deduction.

TIP 3

Know Your Calculator

SUGGESTIONS

- *Which calculator should you use?* The two calculators used for this book are the TI-84 and the TI-Nspire. Both are very powerful. The TI-84 is somewhat easier to use for the functions needed for this test. The TI-Nspire has more features for courses in the future. The choice is up to you. This author prefers the TI-84 for the Algebra Regents. Graphing calculators come with manuals that are as thick as the book you are holding. There are also plenty of video tutorials online for learning how to use advanced features of the calculator. To become an expert user, watch the online tutorials or read the manual.

- *Clearing the memory.* You may be asked at the beginning of the test to clear the memory of your calculator. When practicing for the test, you should clear the memory too so you are practicing under test-taking conditions.

    This is how you clear the memory.

For the TI-84:

Press [2ND] and then [+] to get to the MEMORY menu. Then press [7] for Reset.

```
MEMORY
1:About
2:Mem Mgmt/Del…
3:Clear Entries
4:ClrAllLists
5:Archive
6:UnArchive
7:Reset…
```

Use the arrows to go to [ALL] for All Memory. Then press [1].

```
RAM ARCHIVE ALL
1:All Memory...
```

Press [2] for Reset.

```
RESET MEMORY
1:No
2:Reset
Resetting ALL
will delete all
data, programs &
Apps from RAM &
Archive.
```

The calculator will be reset as if in brand new condition. The one setting that you may need to change is to turn the diagnostics on if you need to calculate the correlation coefficient.

For the TI-Nspire:

The TI-Nspire must be set to Press-To-Test mode when taking the Algebra Regents. Turn the calculator off by pressing [ctrl] and [home]. Press and hold [esc] and then press [home].

```
Press-to-Test
Prevent access to 3D graphing functionality
and pre-existing Scratchpad data, documents
and folders.
   Angle Settings: Degree
   Select additional restrictions:
   ☑ Limit geometry functions
   ☑ Disable function grab and move
  ⦰               Enter Press-to-Test  Cancel
```

While in Press-to-Test mode, certain features will be deactivated. A small green light will blink on the calculator so a proctor can verify the calculator is in Press-to-Test mode.

To exit Press-to-Test mode, use a USB cable to connect the calculator to another TI-Nspire. Then from the home screen on the calculator in Press-to-Test mode, press [doc], [9] and select Exit Press-to-Test.

- *Use parentheses*

  The calculator always uses the order of operations where multiplication and division happen before addition and subtraction. Sometimes, though, you may want the calculator to do the operations in a different order.

  Suppose at the end of a quadratic equation, you have to round $x = \dfrac{-1 + \sqrt{5}}{2}$ to the nearest hundredth. If you enter (−) (1) (+) (2ND) (x^2) (5) (/) (2), it displays

  which is not the correct answer.

  One reason is that for the TI-84 there needs to be a closing parenthesis (or on the TI-Nspire, press [right arrow] to move out from under the radical sign) after the 5 in the square root symbol. Without it, it calculated $-1 + \sqrt{\dfrac{5}{2}}$. More needs to be done, though, since

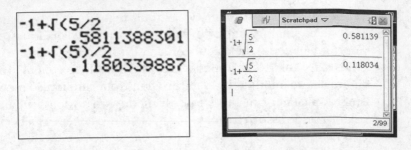

still is not correct. This is the solution to $-1+\sqrt{\dfrac{5}{2}}$.

To get this correct, there also needs to be parentheses around the entire numerator, $-1 + \sqrt{5}$.

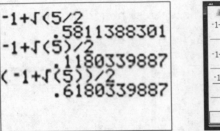

This is the correct answer.

On the TI-Nspire, fractions like this can also be done with [templates].

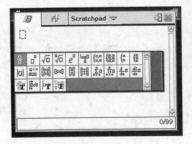

- *Using the ANS feature*
  The last number calculated with the calculator is stored in something called the ANS variable. This ANS variable will appear if you start an expression with a +, −, ×, or ÷. When an answer has a lot of digits in it, this saves time and is also more accurate.

  If for some step in a problem you need to calculate the decimal equivalent of $\frac{1}{7}$, it will look like this on the TI-84:

For the TI-Nspire, if you try the same thing, it leaves the answer as $\frac{1}{7}$. To get the decimal approximation, press [ctrl] and [enter] instead of just [enter].

Now if you want to multiply this by 3, just press [×], and the calculator will display "Ans*"; press [3] and [enter].

	Scratchpad ▽			Scratchpad ▽	
$\frac{1}{7}$		$\frac{1}{7}$	$\frac{1}{7}$		$\frac{1}{7}$
$\frac{1}{7}$		0.142857	$\frac{1}{7}$		0.142857
Ans·3			0.14285714285714·3		0.428571
		2/99			3/99

The ANS variable can also help you do calculations in stages.
To calculate $x = \dfrac{-1 + \sqrt{5}}{2}$ without using so many parentheses
as before, it can be done by first calculating $-1 + \sqrt{5}$ and then
pressing [÷] and [2] and Ans will appear automatically.

```
-1+√(5
          1.236067977
Ans/2
          .6180339887
```

	Scratchpad ▽			Scratchpad ▽	
-1+√5		1.23607	-1+√5		1.23607
Ans/2			$\dfrac{1.2360679774998}{2}$		0.618034
		1/99			2/99

The ANS variable can also be accessed by pressing [2ND] and
[–] at the bottom right of the calculator. If after calculating the
decimal equivalent of 1/7 you wanted to subtract $\dfrac{1}{7}$ from 5,
for the TI-84 press [5], [–], [2ND], [ANS], and [ENTER]. For
the TI-Nspire press [5], [–], [ctrl], [ans], and [enter].

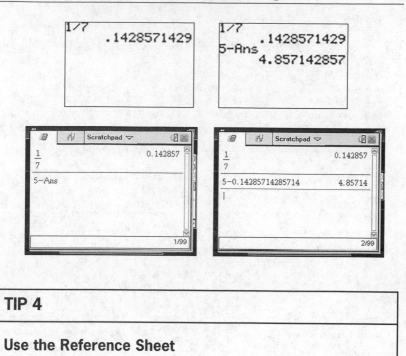

TIP 4

Use the Reference Sheet

SUGGESTIONS

- In the back of the Algebra Regents booklet is a reference sheet that contains 17 conversion facts, such as inches to centimeters and quarts to pints, and also 17 formulas. Many of these conversion facts and formulas will not be needed for an individual test, but the quadratic formula and the arithmetic sequence formula are the two that will come in the handiest.

High School Math Reference Sheet

1 inch = 2.54 centimeters   1 kilometer = 0.62 mile   1 cup = 8 fluid ounces
1 meter = 39.37 inches   1 pound = 16 ounces   1 pint = 2 cups
1 mile = 5280 feet   1 pound = 0.454 kilogram   1 quart = 2 pints
1 mile = 1760 yards   1 kilogram = 2.2 pounds   1 gallon = 4 quarts
1 mile = 1.609 kilometers   1 ton = 2000 pounds   1 gallon = 3.785 liters
   1 liter = 0.264 gallon
   1 liter = 1000 cubic centimeters

Triangle	$A = \frac{1}{2}bh$	Pythagorean Theorem	$a^2 + b^2 = c^2$
Parallelogram	$A = bh$	Quadratic Formula	$x = \frac{-b \pm \sqrt{b^2 - 4ac}}{2a}$
Circle	$A = \pi r^2$	Arithmetic Sequence	$a_n = a_1 + (n-1)d$
Circle	$C = \pi d$ or $C = 2\pi r$	Geometric Sequence	$a_n = a_1 r^{n-1}$
General Prisms	$V = Bh$	Geometric Series	$S_n = \frac{a_1 - a_1 r^n}{1 - r}$ where $r \neq 1$
Cylinder	$V = \pi r^2 h$	Radians	1 radian = $\frac{180}{\pi}$ degrees
Sphere	$V = \frac{4}{3}\pi r^3$	Degrees	1 degree = $\frac{\pi}{180}$ radians
Cone	$V = \frac{1}{3}\pi r^2 h$	Exponential Growth/Decay	$A = A_0 e^{k(t - t_0)} + B_0$
Pyramid	$V = \frac{1}{3}Bh$		

- *How Many Points Do You Need to Pass?*
  The Algebra Regents exam is scored out of a possible 86 points.
  Unlike most tests given in the year by your teacher, the score is
  not then turned into a percent out of 86. Instead each test has
  a conversion sheet that varies from year to year. For example,
  for the June 2014 test, the conversion sheet looked like this.

Raw Score	Scale Score	Raw Score	Scale Score	Raw Score	Scale Score
86	100	57	75	28	64
85	99	56	74	27	63
84	97	55	74	26	62
83	96	54	74	25	61
82	95	53	73	24	60
81	94	52	73	23	59
80	92	51	73	22	58
79	91	50	72	21	56
78	90	49	72	20	55
77	89	48	72	19	54
76	88	47	72	18	52
75	87	46	71	17	50
74	86	45	71	16	49
73	85	44	71	15	47
72	84	43	70	14	45
71	83	42	70	13	42
70	82	41	70	12	40
69	82	40	70	11	38
68	81	39	69	10	35
67	80	38	69	9	32
66	79	37	69	8	30
65	79	36	68	7	26
64	78	35	68	6	23
63	78	34	67	5	20
62	77	33	67	4	16
61	77	32	66	3	12
60	76	31	66	2	9
59	76	30	65	1	4
58	75	29	64	0	0

On this test, 30 points became a 65, 57 points became a 75, and 73 points became an 85. This means that for this examination a student who got 30 out of 86, which is just 35% of the possible points, would get a 65 on this exam. 57 out of 86 is 66%, but this scaled to a 75. 73 out of 86, however, is actually 85% and became an 85. So in the past there has been a curve on the exam for lower scores, though the scaling is not released until after the exam.

Some Key Algebra I Facts and Skills

1. PROPERTIES OF ALGEBRA AND SOLVING LINEAR EQUATIONS WITH ALGEBRA

1.1 ONE-STEP ALGEBRA PROBLEMS

An algebra problem, like $x - 5 = 2$, is one that can be solved by changing both sides of the equation until the variable x is isolated. There are four main properties that can be used in solving algebra problems.

- The **addition property** of equality
  The equation $x - 5 = 2$ is solved by adding 5 to both sides of the equation. When you add to both sides of an equation, you are using the addition property of equality.

  $$x - 5 = 2 \quad \text{the given equation}$$
  $$+5 = +5 \quad \text{addition property of equality}$$
  $$x = 7 \quad x \text{ is isolated. The solution is 7.}$$

- The **subtraction property** of equality
  The equation $x + 2 = 7$ is solved by subtracting 2 from both sides of the equation. When you subtract from both sides of an equation, you are using the subtraction property of equality.

  $$x + 2 = 7 \quad \text{the given equation}$$
  $$-2 = -2 \quad \text{subtraction property of equality}$$
  $$x = 5 \quad x \text{ is isolated. The solution is 5.}$$

- The **division** property of equality

  The equation $2x = 10$ is solved by dividing both sides of the equation by 2. When you divide both sides of the equation, you are using the division property of equality.

$$2x = 10 \quad \text{the given equation}$$

$$\frac{2x}{2} = \frac{10}{2} \quad \text{division property of equality}$$

$$x = 5 \quad x \text{ is isolated. The solution is 5.}$$

- The **multiplication** property of equality

  The equations $\frac{x}{5} = 3$ and $\frac{2}{3}x = 8$ can be solved by multiplying both sides of the equation by the same number.

$$\frac{x}{5} = 3 \qquad \text{the given equation}$$

$$5 \cdot \frac{x}{5} = 5 \cdot 3 \qquad \text{multiplication property of equality}$$

$$x = 15 \qquad x \text{ is isolated. The solution is 15.}$$

$$\frac{2}{3}x = 8 \qquad \text{the given equation}$$

$$\frac{3}{2} \cdot \frac{2}{3}x = \frac{3}{2} \cdot 8 \qquad \text{multiplication property of equality}$$

$$x = 12 \qquad x \text{ is isolated. The solution is 12.}$$

1.2 TWO-STEP ALGEBRA PROBLEMS

When an equation has the form $mx + b = y$, it takes two steps to solve for x. The first step is to eliminate the b, which is called the **constant**. The second step is to eliminate the m, which is called the **coefficient**. The b is eliminated with either the addition or the subtraction property of equality. The m is eliminated with either the division or the multiplication property of equality.

- For the equation $3x - 7 = 11$

$$3x - 7 = 11 \qquad \text{the given equation}$$
$$+7 = +7 \qquad \text{addition property of equality}$$
$$\frac{3x}{3} = \frac{18}{3} \qquad \text{division property of equality}$$
$$x = 6 \qquad x \text{ is isolated. The solution is 6.}$$

$$\frac{2}{3}x + 5 = 11 \qquad \text{the given equation}$$
$$-5 = -5 \qquad \text{subtraction property of equality}$$
$$\frac{2}{3}x - 6$$

$$\frac{3}{2} \cdot \frac{2}{3}x = \frac{3}{2} \cdot 6 \qquad \text{multiplication property of equality}$$
$$x = 9 \qquad x \text{ is isolated. The solution is 9.}$$

1.3 COMBINING LIKE TERMS BEFORE SOLVING

If the equation has multiple x terms or multiple constants, the equation should first be simplified by combining like terms. After all like terms have been combined, the question will usually be a two-step algebra problem and can be solved with the methods from Section 1.2.

$$2x + 5 + 3x - 2 = 23 \qquad \text{the given equation}$$

$$2x + 3x + 5 - 2 = 23 \qquad \text{Terms are rearranged so the } x \text{ terms are together and the constants are together. This step is optional.}$$

$$5x + 3 = 23 \qquad \text{Like terms have been combined.}$$

$$-3 = -3 \qquad \text{subtraction property of equality}$$

$$\frac{5x}{5} = \frac{20}{5} \qquad \text{division property of equality}$$

$$x = 4 \qquad x \text{ is isolated. The solution is 4.}$$

1.4 VARIABLES ON BOTH SIDES OF THE EQUATION

When there are x terms on both sides of the equation, the addition property of equality or the subtraction property of equality can be used to change the equation into one where the x terms are all on the same side of the equation.

$5x - 3 = 12 + 2x$	the given equation
$-2x = -2x$	subtraction property of equality eliminates the x term from the right-hand side of the equation
$3x - 3 = 12$	
$+3 = +3$	addition property of equality
$\dfrac{3x}{3} = \dfrac{15}{3}$	division property of equality
$x = 5$	x is isolated. The solution is 5.

1.5 EQUATIONS WITH MORE THAN ONE VARIABLE

An equation with more than one variable can be solved the same way as an equation with just one variable. The **solution** will not be a number in these problems, but an **expression** with variables and numbers in it.

$ax + 2 = c$	the given equation
$-2 = -2$	subtraction property of equality
$ax = c - 2$	the c and the 2 cannot be combined since they are unlike terms
$\dfrac{ax}{a} = \dfrac{c-2}{a}$	division property of equality
$x = \dfrac{c-2}{a}$	x is isolated. The solution for x is not a number but an expression in terms of c and a. The answer is $\dfrac{c-2}{a}$.

Practice Exercises

1. Antonio started the question $2x + 1 = 11$ by writing $2x = 10$. Which property justifies this step?
   (1) Commutative property of addition
   (2) Distributive property of multiplication over addition
   (3) Addition property of equality
   (4) Subtraction property of equality

2. Mila used the multiplication property to justify the first step in solving an equation. The original equation was $\frac{x}{2} + 4 = 10$. What could the equation have been transformed into after this step?

   (1) $\frac{x}{2} = 6$ (3) $x + 8 = 20$

   (2) $\frac{x}{2} + 6 + 12$ (4) $x + 2 = 5$

3. What is the solution set for the equation $x - 6 = 7$?
   (1) {1}   (3) {11}
   (2) {6}   (4) {13}

4. What value of x makes the equation $3x + 7 = 22$ true?
   (1) 1   (3) 5
   (2) 3   (4) 7

5. Find the solution set for the equation $5(x + 4) = 35$.

(1) $\{1\}$ (3) $\{3\}$

(2) $\{2\}$ (4) $\{4\}$

6. Solve for d in terms of $c, e,$ and f.

$cd - e = f.$

(1) $\dfrac{f - e}{c}$ (3) $\dfrac{f}{c} + e$

(2) $\dfrac{f + e}{c}$ (4) $\dfrac{f}{c} - e$

7. Solve for m in terms of $a, b,$ and c.

$b - ma = c.$

(1) $\dfrac{b - c}{-a}$ (3) $\dfrac{c - b}{-a}$

(2) $(c - b) - a$ (4) $(c + b) - a$

8. Solve for r in terms of c and π.

$c = 2\pi r$

(1) $\dfrac{2c}{\pi}$ (3) $\dfrac{c}{2\pi}$

(2) $\dfrac{2\pi}{c}$ (4) $\dfrac{2}{c\pi}$

Solutions

1. The first step of the process is to subtract 1 from each side of the equation. This is called the subtraction property of equality. The correct choice is **(4)**.

2. Though it is more common to begin this question by subtracting 4 from both sides of the equation, in this case she does it by multiplying both sides of the equation by 2. The left-hand side becomes $x + 8$ and the right-hand side becomes 20. The correct choice is **(3)**.

3. Isolate the x by adding 6 to both sides of the equation. The equation then becomes $x = 13$. The correct choice is **(4)**.

4. Subtract 7 from both sides of the equation to get $3x = 15$. Divide both sides of the equation by 3 to get $x = 5$. The correct choice is **(3)**.

5. One way to solve this equation is to first distribute the 5 through the left-hand side to get the equation $5x + 20 = 35$, then subtract 20 from both sides of the equation to get $5x = 15$, and finally to divide both sides of the equation by 5 to get $x = 3$. Another way is to first divide both sides of the equation by 5 to get $x + 4 = 7$, and then subtract 4 from both sides of the equation to get $x = 3$. The correct choice is **(3)**.

6. First add e to both sides of the equation to get $cd = f + e$. Then divide both sides of the equation by c to get $d = (f + e)/c$. The correct choice is **(2)**.

7. First subtract b from both sides of the equation to get $-ma = c - b$. Then divide both sides by $-a$ to get $m = \dfrac{c-b}{-a}$. The correct choice is **(3)**.

8. Divide both sides of the equation by 2π to get $\dfrac{c}{2\pi} = r$. The correct choice is **(3)**.

2. POLYNOMIAL ARITHMETIC

2.1 CLASSIFYING POLYNOMIALS

A **polynomial** is an expression like $2x + 5$ or $3x^2 - 5x + 3$. The **terms** of a polynomial are separated by + or – signs. The polynomial $2x + 5$ has two terms. The polynomial $3x^2 - 5x + 3$ has three terms. The terms of a polynomial have a **coefficient** and a **variable part**. The term $3x^2$ has a coefficient of 3 and a variable part of x^2. A term with no variable part is called a **constant**.

- A polynomial with three terms is called a **trinomial**.
- A polynomial with two terms is called a **binomial**.
- A polynomial with one term is called a **monomial**.

2.2 MULTIPLYING AND DIVIDING MONOMIALS

- To *multiply* one monomial by another, multiply the coefficients and multiply the variable parts by adding the exponents on the same variables.

$$8x^3 \cdot 2x$$

Multiply the coefficients $8 \cdot 2 = 16$.
Multiply the variable parts by adding the exponents $x^3 \cdot x^1 = x^4$.
The solution is $16x^4$.

- To *divide* one monomial by another, divide the coefficients and divide the variable parts by subtracting the exponents on the same variables.

$$8x^3 \div 2x$$

Divide the coefficients. $8 \div 2 = 4$.
Divide the variable parts by subtracting the exponents

$$x^3 \div x^1 = x^2$$

The solution is $4x^2$.

This question can also be expressed as $\dfrac{8x^3}{2x} = 4x^2$

2.3 COMBINING LIKE TERMS

Like terms are terms that have the same variable part. For example, $3x^2$ and $2x^2$ are like terms because the variable part for both is x^2. Like terms can be added or subtracted by adding or subtracting the coefficients and by not changing the variable part.

$$3x^2 + 2x^2 = 5x^2$$
$$3x^2 - x^2 = 3x^2 - 1x^2 = 2x^2$$

- To simplify $2x + 3 + 4x - 5$, combine the like terms with the variable part of x. $2x + 4x = 6x$. Also combine the constants $3 - 5 = -2$. This expression simplifies to $6x - 2$.

If terms are not like terms, they cannot be combined by adding or subtracting. $3x^2 + 5x^3$ cannot be combined because the exponents are different, so they are not like terms.

2.4 MULTIPLYING MONOMIALS AND POLYNOMIALS

- To *multiply* a polynomial by a monomial, use the **distributive property**.

$$2(3x + 5) = 2 \cdot 3x + 2 \cdot 5 = 6x + 10$$

This works for more complex monomials and polynomials also.

$$2x^2(5x^2 - 7x + 3) = 2x^2 \cdot 5x^2 + 2x^2 \cdot -7x + 2x^2 \cdot 3$$
$$= 10x^4 - 14x^3 + 6x^2$$

2.5 ADDING AND SUBTRACTING POLYNOMIALS

- To *add* two polynomials, remove the parentheses from both and combine like terms.

$$(5x + 2) + (3x - 4) = 5x + 2 + 3x - 4 = 8x - 2$$

- To **subtract** two polynomials, remove the parentheses of the polynomial on the left, then negate all the terms of the polynomial on the right, and remove the parentheses before combining like terms.

$$(5x + 2) - (3x - 4) = 5x + 2 - 3x + 4 = 2x + 6$$

2.6 MULTIPLYING BINOMIALS

- To *multiply* binomials, use the <u>FOIL</u> process.

$$(2x + 3)(5x - 2)$$

The F stands for firsts. Multiply $2x \cdot 5x$, the first term in each of the parentheses. $10x^2$.

The O stands for outers. Multiply $2x \cdot -2$, the terms on the far left and on the far right. $-4x$.

The I stands for inners. Multiply $3 \cdot 5x$, the two terms in the middle. $15x$.

The L stands for lasts. Multiply $3 \cdot -2$, the second term in each of the parentheses. -6

These four answers become $10x^2 - 4x + 15x - 6$. Combine like terms to get $10x^2 + 11x - 6$ in simplified form.

2.7 FACTORING POLYNOMIALS

Factoring a number, like 15, is when two numbers are found that can be multiplied to become that number, $15 = 3 \cdot 5$. Factoring polynomials is more involved and there are certain patterns to be aware of.

- **Greatest Common Factor Factoring**

The terms of some polynomials have a greatest common factor. This can be factored out like a reverse use of the distributive property.

In $6x^2 + 8x$, the terms have a common factor of $2x$. Write $2x$ outside the parentheses and divide each term by $2x$ to determine what goes inside the parentheses.

$$2x(3x + 4)$$

- **Difference of Perfect Squares Factoring**

The expression $a^2 - b^2$ can be factored into $(a - b)(a + b)$. This works anytime both terms of a binomial are perfect squares and there is a minus sign between the two terms.

$$x^2 - 9 = x^2 - 3^2 = (x - 3)(x + 3)$$

- **Reverse FOIL**

A **trinomial** like $x^2 + 8x + 15$ can be factored if there are two numbers that have a sum equal to the coefficient of the x term, 8, and a product equal to the constant 15. Since $3 + 5 = 8$ and $3 \cdot 5 = 15$,

$$x^2 + 8x + 15 = (x + 3)(x + 5).$$

For $x^2 + 3x - 10$, the numbers that have a sum of 3 and a product of -10 are -2 and 5.

$$x^2 + 3x - 10 = (x - 2)(x + 5)$$

2.8 MORE COMPLICATED FACTORING

Sometimes none of the factoring patterns seems to match the polynomial that needs to be factored. When this happens, see if it is possible to rewrite it in a way that resembles the pattern better.

- The polynomial $x^4 - 9$ can be rewritten as $(x^2)^2 - 3^2$, which now has the difference of perfect squares pattern.

$$x^4 - 9 = (x^2)^2 - 3^2 = (x^2 - 3)(x^2 + 3)$$

- The polynomial $x^4 + 8x^2 + 15$ can be rewritten as

$$(x^2)^2 + 8(x^2) + 15.$$

$$x^4 + 8x^2 + 15 = (x^2)^2 + 8(x^2) + 15 = (x^2 + 3)(x^2 + 5)$$

Practice Exercises

1. Classify this polynomial $5x^2 + 3$.
    (1) Monomial
    (2) Binomial
    (3) Trinomial
    (4) None of the above

2. Classify this polynomial $7x^2 - 3x + 2$.
    (1) Monomial
    (2) Binomial
    (3) Trinomial
    (4) None of the above

3. Multiply $3x^3 \cdot 4x^5$.
    (1) $7x^8$
    (2) $7x^{15}$
    (3) $12x^8$
    (4) $12x^{15}$

4. Which expression is equivalent to $2x^2 + 5x^2$?
    (1) $10x^2$
    (2) $7x^2$
    (3) $7x^4$
    (4) The expression cannot be simplified any further.

5. Simplify $6x(2x + 3)$.
    (1) $8x^2 + 18x$
    (2) $12x^2 + 18x$
    (3) $12x^2 + 3$
    (4) $20x$

6. Simplify $3x(5x^2 - 2x + 3)$.
   (1) $15x^3 - 2x + 3$ (3) $5x^2 + x + 3$
   (2) $15x^3 - 6x^2 + 9x$ (4) $15x^3 + 6x^2 - 9x$

$3x - 4 - 5x + 3$

7. Simplify $(3x - 4) - (5x - 3)$.
   (1) $-2x - 7$ (3) $2x - 1$
   (2) $-2x - 1$ (4) $2x - 7$

$6x^2 - 2x + 9x = 3$

8. $(2x + 3)(3x - 1) =$
   (1) $6x^2 - 3$ (3) $6x^2 + 7x - 3$
   (2) $6x^2 + 11x - 3$ (4) $6x^2 - 7x - 3$

9. Factor $x^2 - 2x - 15$.
   (1) $(x - 3)(x - 5)$ (3) $(x - 15)(x + 1)$
   (2) $(x + 3)(x - 5)$ (4) $(x + 15)(x - 1)$

10. Factor $x^4 + 7x^2 + 12$.
   (1) $(x^2 + 6)(x^2 + 2)$
   (2) $(x^2 - 3)(x^2 - 4)$
   (3) $(x^2 + 3)(x^2 + 4)$
   (4) This cannot be factored.

Solutions

1. There are two terms, $5x^2$ and 3 separated by a + sign. A polynomial with two terms is called a binomial. The correct choice is (**2**).

2. There are three terms, $7x^2$, $3x$, and 2 separated by + and − signs. A polynomial with three terms is called a trinomial. The correct choice is (**3**).

3. To multiply two monomials, first multiply the coefficients, $3 \cdot 4 = 12$. Then multiply the variable parts. Remember that when multiplying variables, you add the exponents. $x^3 \cdot x^5 = x^{(3+5)} = x^8$. The solution is $12x^8$. The correct choice is (**3**).

4. Since these are like terms with variable part x^2, they can be combined. The solution will also have a variable part of x^2 with a coefficient equal to the sum of the two coefficients. Since $2 + 5 = 7$, $2x^2 + 5x^2 = 7x^2$. The correct choice is (**2**).

5. Using the distributive property it becomes $6x \cdot 2x + 6x \cdot 3 = 12x^2 + 18x$. The correct choice is (**2**).

6. Using the distributive property it becomes $3x \cdot 5x^2 + 3x(-2x) + 3x(3) = 15x^3 - 6x^2 + 9x$. The correct choice is (**2**).

7. Distribute the − sign through the parentheses on the right. The expression becomes $3x - 4 - 5x + 3$. Combine like terms to get $-2x - 1$. The correct choice is (**2**).

8. Use the FOIL process. The firsts are $2x \cdot 3x = 6x^2$. The outers are $2x \cdot (-1) = -2x$. The inners are $3 \cdot 3x = 9x$. The lasts are $3 \cdot (-1) = -3$. Combine these four terms to get $6x^2 - 2x + 9x - 3$. Combine like terms to get $6x^2 + 7x - 3$. The correct choice is (**3**).

9. To factor this quadratic trinomial, find two numbers that have a product of −15 and a sum of −2. The numbers are −5 and +3. The factors, then, are $(x - 5)(x + 3)$. The correct choice is (**2**).

10. This trinomial can be written as $(x^2)^2 + 7(x^2) + 12$. It has the same structure, then, as a quadratic trinomial and can be factored by finding two numbers that have a product of 12 and a sum of 7. The two numbers are $+3$ and $+4$. The factors are $(x^2 + 3)$ and $(x^2 + 4)$. The correct choice is **(3)**.

3. QUADRATIC EQUATIONS

3.1 METHODS OF SOLVING QUADRATIC EQUATIONS

A **quadratic equation** is an equation that can be written in the form $ax^2 + bx + c = 0$. For example, $x^2 + 4x - 5 = 0$ is a quadratic equation. There are three ways to solve a quadratic equation.

- 1. Solve by factoring. If possible, factor the left-hand side of the equation.

$$x^2 + 4x - 5 = 0$$
$$(x + 5)(x - 1) = 0$$

Since the only way that two things can have a product of zero is if at least one of them is zero, this means that either $(x + 5)$ or $(x - 1)$ must equal zero.

$$x + 5 = 0 \text{ or } x - 1 = 0$$
$$-5 = -5 \quad\quad +1 = +1$$
$$x = -5 \text{ or } \quad x = 1$$

$x = -5$ or $x = 1$ are the solutions to this quadratic equation.

- 2. Solve by completing the square. First eliminate the constant from the left-hand side by adding or subtracting.

$$x^2 + 4x - 5 = 0$$
$$+ 5 = + 5$$
$$x^2 + 4x = 5$$

Next, *divide* the coefficient of the x by 2, square that answer, and add that number to both sides of the equation.

$$\left(\frac{4}{2}\right)^2 = 2^2 = 4$$

$$x^2 + 4x + 4 = 5 + 4$$
$$x^2 + 4x + 4 = 9$$

The left-hand side of the equation will factor.

$$(x + 2)(x + 2) = 9$$
$$(x + 2)^2 = 9$$

- Take the square root of both sides of the equation, putting a ± in front of the square root of the right-hand side.

$$\sqrt{(x+2)^2} = \pm\sqrt{9}$$
$$x + 2 = \pm 3$$
$$-2 = -2$$
$$x = -2 \pm 3$$
$$x = -2 + 3 \text{ or } x = -2 - 3$$
$$x = 1 \text{ or } x = -5$$

- 3. Solve with the quadratic formula. Any quadratic equation of the form $ax^2 + bx + c = 0$ can be solved with the equation

$$x = \frac{-b \pm \sqrt{b^2 - 4ac}}{2a}.$$

For $x^2 + 4x - 5 = 0$, $a = 1$, $b = 4$, $c = -5$.

$$x = \frac{-4 \pm \sqrt{4^2 - 4\cdot 1\cdot(-5)}}{2\cdot 1} = \frac{-4 \pm \sqrt{16+20}}{2} = \frac{-4 \pm \sqrt{36}}{2} = \frac{-4 \pm 6}{2}$$

$$x = \frac{-4+6}{2} = \frac{2}{2} = 1 \text{ or } x = \frac{-4-6}{2} = -\frac{10}{2} = -5$$

3.2 THE RELATIONSHIP BETWEEN FACTORS AND ZEROS

If a quadratic equation has factors $(x - p)$ and $(x - q)$ then the roots of the equation are p and q. If the equation has roots (or zeros) p and q, the factors are $(x - p)$ and $(x - q)$.

For example, the factors of the equation $x^2 - 7x + 10$ are $(x - 5)$ and $(x - 2)$. Therefore, the zeros of the equation are 5 and 2.

If the roots of a quadratic equation are -2 and 8, then the factors are $(x - (-2))$ and $(x - 8)$. The $(x - (-2))$ can be expressed as $(x + 2)$.

3.3 WORD PROBLEMS INVOLVING QUADRATIC EQUATIONS

Some real-world scenarios can be modeled with quadratic equations.

- **Area Problems**

The width of a rectangle is 3 units more than the length. If the area of the rectangle is 70 square units, what are the length and width of the rectangle?

$$A = 70 \quad l + 3$$
$$l$$

Since area is length times width, for this scenario

$$70 = l \cdot w$$
$$70 = l \cdot (l + 3)$$
$$70 = l^2 + 3l$$
$$0 = l^2 + 3l - 70$$

Any of the methods can be used to solve for the answers $l = -10$ and $l = 7$. Since the length must be positive, the answer is 7 units.

- **Projectile Problems**

The height of a projectile after t seconds can be modeled with a quadratic equation. If the equation for the height of a baseball is $h = 16t^2 + 48t + 64$, when will the baseball land on the ground?

When the ball lands on the ground, the height will be 0.

$0 = -16t^2 + 48t + 64$ is the equation.

- This can be solved by any of the methods for the answers $t = -1$ and $t = 4$. Since the amount of time must be positive, the solution is 4 seconds.

Practice Exercises

1. Find all solutions to $(x + 2)^2 = 64$.
   (1) $\sqrt{62}$, $-\sqrt{62}$ (3) 6
   (2) 6, –10   (4) –10

2. Use completing the square to find both solutions for x in the
   equation $x^2 + 8x + 16 = 9$.

   $x^2 + 8x = -7$

   (1) –1, –7   (3) –2, –3
   (2) –1, –2   (4) –3, –4

3. Find all solutions to $x^2 + 6x = 0$.
   (1) 0   (3) 0, –6
   (2) –6   (4) 0, 6

4. Solve $x^2 + 10x + 24 = 0$ by factoring.
   (1) 4, –6   (3) –4, 6
   (2) –4, –6   (4) 4, 6

5. If the roots of an equation are 3 and –6, what could the equation be?
   (1) $(x - 3)(x - 6) = 0$ (3) $(x + 3)(x - 6) = 0$
   (2) $(x - 3)(x + 6) = 0$ (4) $(x + 3)(x + 6) = 0$

6. If the roots of a polynomial are 1 and –8, what could be the factors?
   (1) $(x - 1)$ and $(x + 8)$ (3) $(x + 1)$ and $(x + 8)$
   (2) $(x - 1)$ and $(x - 8)$ (4) $(x + 1)$ and $(x - 8)$

7. Solve $x^2 + 4x - 7 = 0$ with the quadratic formula.
   (1) 1.3   (3) $2 \pm \sqrt{12}$
   (2) $2 \pm \sqrt{11}$ (4) $-2 \pm \sqrt{11}$

8. The width of a rectangle is 10 inches longer than its length. If the area of the rectangle is 56 square inches, which equation could be used to determine its length (l)?
   (1) $l(l + 10) = 56$ (3) $2l + 2(l + 10) = 56$
   (2) $l(l - 10) = 56$ (4) $2l - 2(l + 10) = 56$

9. The height of a projectile in feet at time t is determined by the equation $h = -16t^2 + 128t + 320$. At what time will the projectile be 560 feet high?
   (1) 4 seconds   (3) 6 seconds
   (2) 5 seconds   (4) 7 seconds

10. The height of a projectile in feet at time t is determined by the equation $h = -16t^2 + 112t + 128$. At what time will the projectile be 0 feet high?
    (1) 5 seconds   (3) 7 seconds
    (2) 6 seconds   (4) 8 seconds

Solutions

1. Take the square root of both sides to get $x + 2 = \pm 8$. The equation $x + 2 = 8$ has solution $x = 6$. The equation $x + 2 = -8$ has solution $x = -10$. The correct choice is (**2**).

2. Since the constant, 16, is already equal to the square of half the coefficient $\left(\dfrac{8}{2}\right)^2$, the left-hand side of the equation is already a perfect square trinomial. It can be factored as $(x + 4)^2 = 9$. Then take the square root of both sides to get $x + 4 = \pm 3$. The equation $x + 4 = 3$ has solution $x = -1$. The equation $x + 4 = -3$ has solution $x = -7$. The correct choice is (**1**).

3. Factor out an x to get the equation $x(x + 6) = 0$. This equation is true when either $x = 0$ or when $x + 6 = 0$, which becomes $x = -6$. The correct choice is (**3**).

4. The two numbers that have a product of 24 and a sum of 10 are $+4$ and $+6$. The factors are $(x + 4)(x + 6)$. The solutions to the equation $(x + 4)(x + 6) = 0$ are when $x + 4 = 0$, which becomes $x = -4$ and also when $x + 6 = 0$, which becomes $x = -6$. The correct choice is (**2**).

5. When a is a root of an equation, $(x - a)$ is a factor. If the roots are 3 and -6, the factors can be $(x - 3)$ and $(x - (-6)) = (x + 6)$. The correct choice is (**2**).

6. If 1 is a root, $(x - 1)$ is a factor. If -8 is a root $(x - (-8)) = (x + 8)$ is a factor. The correct choice is (**1**).

7. $a = 1, b = 4, c = -7$.

$$x = \frac{-4 \pm \sqrt{4^2 - 4(1)(-7)}}{2} = \frac{-4 \pm \sqrt{16 + 28}}{2} = \frac{-4 \pm \sqrt{44}}{2}$$

$$= \frac{-4 \pm 2\sqrt{11}}{2} = -2 \pm \sqrt{11}.$$

The correct choice is (**4**).

8. The area of a rectangle is $l \cdot w$. If l is the length and $l + 10$ is the width, the area is $l(l + 10)$. If the area is known to be 56, the equation that can be used to solve for l is $l(l + 10) = 56$. The correct choice is **(1)**.

9. When 560 is substituted for h, the equation becomes $560 = -16t^2 + 128t + 320$. Subtract 560 from both sides of the equation to get $0 = -16t^2 + 128t - 240$. Divide both sides by -16 to get $0 = t^2 - 8t + 15$. The right-hand side factors and the equation becomes $0 = (t - 3)(t - 5)$ with solutions $t = 3$ and $t = 5$. Of the choices listed, only 5 seconds is correct. The correct choice is **(2)**.

10. Substituting $h = 0$ into the equation, it becomes $0 = -16t^2 + 112t + 128$. Divide both sides by -16 to get $0 = t^2 - 7t - 8$. The right side factors so the equation becomes $0 = (t - 8)(t + 1)$. The solutions are $t = 8$ and $t = -1$. Since the amount of time must be positive, the -1 is rejected. The correct choice is **(4)**.

4. SYSTEMS OF LINEAR EQUATIONS

4.1 WHAT IS A SYSTEM OF LINEAR EQUATIONS?

A **system of equations** is a set of two equations that each have two variables. The system is *linear* if there are no exponents greater than one on any of the variables.

$$2x + 3y = 21$$
$$5x - 2y = 5$$

is a system of linear equations.

The solution to a system of equations is the set of **ordered pairs** that satisfy both equations at the same time. For the system above, the solution is the ordered pair $(3, 5)$ since

$$2(3) + 3(5) = 6 + 15 = 21$$
$$5(3) - 2(5) = 15 - 10 = 5$$

There are two main techniques for solving systems of equations.

4.2 SOLVING A SYSTEM OF EQUATIONS WITH THE SUBSTITUTION METHOD

If one of the two equations is in the form $y = mx + b$, use **the substitution method**.

$$y = 2x + 3$$
$$2x + 4y = 42$$

- Substitute the expression $2x + 3$ for the y in the second equation.

$$2x + 4(2x + 3) = 42$$

Solve for x.

$$2x + 8x + 12 = 42$$
$$10x + 12 = 42$$
$$-12 = -12$$
$$\frac{10x}{10} = \frac{30}{10}$$
$$x = 3$$

- Substitute 3 for x into either of the equations and solve for y.

$$y = 2(3) + 3$$
$$y = 6 + 3$$
$$y = 9$$

The solution is the ordered pair $(3, 9)$.

4.3 SOLVING A SYSTEM OF EQUATIONS WITH THE ELIMINATION METHOD

When both equations are in the form $ax + by = c$, use the **elimination method**. The elimination method is when the two equations are combined in a way that one of the variables is eliminated.

In the system of equations below, the y term will be eliminated if the two equations are added. When the coefficient of one of the variables in one equation is the same number with the opposite sign of the same variable in the other equation, add the equations to eliminate that variable. In this case, the $+2y$ has the opposite coefficient as the $-2y$.

$$
\begin{array}{r}
3x + 2y = 19 \\
+\ 4x - 2y = 16 \\
\hline
\dfrac{7x}{7} = \dfrac{35}{7} \\
x = 5
\end{array}
$$

- Substitute 5 for x into either of the equations

$$3(5) + 2y = 19$$
$$15 + 2y = 19$$
$$-15 = -15$$
$$\frac{2y}{2} = \frac{4}{2}$$
$$y = 2$$

The solution is the ordered pair $(5, 2)$.

For some systems of equations, one or both of the equations need to be changed so the elimination will happen.

In the system

$$2x - 4y = 4$$
$$3x + 2y = 14$$

Since 4 is a multiple of 2, multiply both sides of the bottom equation by 2 and the coefficients of the y variables will be the same number with opposite signs.

$$2x - 4y = 4$$
$$2(3x + 2y) = 2(14)$$

$$2x - 4y = 4$$
$$+ \ 6x + 4y = 28$$

$$\frac{8x}{8} = \frac{32}{8}$$

$$x = 4$$

$$2(4) - 4y = 4$$
$$8 - 4y = 4$$
$$-8 = -8$$

$$\frac{-4y}{-4} = \frac{-4}{-4}$$

$$y = 1$$

The solution is the ordered pair (4, 1).

In the system

$$3x - 5y = 17$$
$$2x + 4y = 4$$

- To eliminate the y, determine the least common multiple of 4 and 5, which is 20. Multiply both sides of both equations so that one of the coefficients on the y is -20 and the other is $+20$. To do this, multiply both sides of the top equation by 4 and multiply both sides of the bottom equation by 5.

$$4(3x - 5y) = 4(17)$$
$$5(2x + 4y) = 5(4)$$

$$\begin{aligned} 12x - 20y &= 68 \\ +\quad 10x + 20y &= 20 \end{aligned}$$

$$\frac{22x}{22} = \frac{88}{22}$$

$$x = 4$$

- Substitute 4 for x in either of the original equations and solve for y.

$$3(4) - 5y = 17$$
$$12 - 5y = 17$$
$$-12 = -12$$

$$\frac{-5y}{-5} = \frac{5}{-5}$$

$$y = -1$$

The solution to the system of equations is the ordered pair $(4, -1)$.

4.4 WORD PROBLEMS INVOLVING SYSTEMS OF EQUATIONS

Some real-world scenarios can be modeled with a system of linear equations. Here is a typical example.

If five slices of pizza and three drinks cost $21 and two slices of pizza and five drinks cost $16, how much is it for just one slice of pizza?

- Let x be the cost of a slice of pizza and y be the cost of a drink.

The system of equations is

$$5x + 3y = 21$$
$$2x + 5y = 16$$

- To eliminate the y, make the $3y$ and the $5y$ into $15y$ and $-15y$, respectively, by multiplying both sides of the top equation by 5 and both sides of the bottom equation by -3. Then add the two equations, and solve for x.

$$5(5x + 3y) = 5\,(21)$$
$$-3(2x + 5y) = -3\,(16)$$

$$25x + 15y = 105$$
$$+\ -6x - 15y = -48$$

$$\frac{19x}{19} = \frac{57}{19}$$
$$x = 3$$

Since the question just asked for the price of a slice of pizza, it is not necessary to also find the value of y. A slice of pizza costs $3.

Practice Exercises

1. $(2, 5)$ is a solution to which equation?
   (1) $x + 2y = 9$ (3) $8x - y = 9$
   (2) $2x + y = 9$ (4) $3x + 4y = 25$

2. Which equation has the same solution set as the equation $2x + 3y = 5$?
   (1) $8x + 12y = 20$ (3) $6x + 9y = 12$
   (2) $8x + 12y = 15$ (4) $4x + 6y = 8$

3. Solve the system of equations
$$3x + 2y = 17$$
$$4x - 2y = 4$$

   (1) $(4, 3)$ (3) $(3, 4)$
   (2) $(5, 1)$ (4) $(1, 6)$

4. Solve the system of equations
$$y = 5x + 3$$
$$2x + 6y = 50$$

   (1) $(1, 8)$ (3) $(-1, -8)$
   (2) $(8, 1)$ (4) $(-8, -1)$

5. Solve the system of equations

$$2x + 3y = -1$$
$$-2x + 5y = -23$$

(1) (−4, 3)   (3) (4, −3)
(2) (4, 3)   (4) (−4, −3)

6. Solve the system of equations

$$y = 3x - 2$$
$$4x - 2y = -4$$

(1) (10, 4)   (3) (−10, −4)
(2) (4, 10)   (4) (−4, −10)

Handwritten:
$$4x - 2(3x - 2)$$
$$4x - 6x + 4 = 44$$
$$-2x = -8$$

7. In order to eliminate the *x* from this system of equations,

$$12x - 3y = 21$$
$$-2x + 6y = 2$$

you could
(1) Multiply both sides of the first equation by 2.
(2) Multiply both sides of the second equation by 6.
(3) Multiply both sides of the first equation by −2.
(4) Multiply both sides of the second equation by 1/2.

8. Solve the system of equations

$$8x - 2y = 28$$
$$4x + 3y = 6$$

(1) (3, 2)   (3) (−3, −2)
(2) (−3, 2)   (4) (3, −2)

Handwritten:
$$8x - 2y = 28$$
$$-8x - 6y = -12$$
$$-8y = 16$$

9. Which system of equations can be used to model the following scenario?

   There are 50 animals. Some of the animals have two legs and the rest of them have four legs. In total there are 172 legs.
   (1) $x + y = 172$
   $2x + 4y = 50$
   (2) $x + 50 = y$
   $2x + 172 = 4y$
   (3) $y + 50 = x$
   $4y + 172 = 2x$
   (4) $x + y = 50$
   $2x + 4y = 172$

10. A pet store has 30 animals. Some are cats and the rest are dogs. The cats cost $50 each. The dogs cost $100 each. If the total cost for all 30 animals is $1,900, how many cats are there?
   (1) 8
   (2) 20
   (3) 22
   (4) 24

Solutions

1. Substitute 2 for x and 5 for y into each of the choices. Choice (1) becomes $2 + 2(5) = 9$, which is not true. Choice (2) becomes $2(2) + 5 = 9$, which is true. The correct choice is **(2)**.

2. If both sides of an equation are multiplied by the same number, the new equation has the same solution set as the original equation. If both sides of the equation $2x + 3y = 5$ are multiplied by 4, it becomes $8x + 12y = 20$, which is choice (1). The other choices could be obtained by multiplying the right-hand side of the original equation by one number and the left-hand side of the original equation by another number, which will not produce an equation with the same solution set as the original. The correct choice is **(1)**.

3. Since the top equation has a $+2y$ and the bottom equation has a $-2y$, the equations can be added together, and the y terms will drop out leading to the equation $7x = 21$ or $x = 3$. To solve for y, substitute 3 for x into either of the original equations, like $3(3) + 2y = 17$, $9 + 2y = 17$, $2y = 8$, which leads to $y = 4$. The correct choice is **(3)**.

4. Since the y is isolated in the top equation, substitute $5x + 3$ for y in the bottom equation. It becomes $2x + 6(5x + 3) = 50$, $2x + 30x + 18 = 50$, $32x + 18 = 50$, $32x = 32$, $x = 1$. Only one choice has an x-coordinate of 1. The correct choice is **(1)**.

5. Since the top equation has a $+2x$ and the bottom equation has a $-2x$, the equations can be added together to get $8y = -24$, $y = -3$. Substitute -3 for y into either equation to solve for x. $2x + 3(-3) = -1$, $2x - 9 = -1$, $2x = 8$, $x = 4$. The correct choice is **(3)**.

6. Since the y is isolated in the top equation, substitute $3x - 2$ for y in the bottom equation. $4x - 2(3x - 2) = -4$, $4x - 6x + 4 = -4$, $-2x + 4 = -4$, $-2x = -8$, $x = 4$. To solve for y, substitute 4 for x into either equation. $y = 3(4) - 2 = 12 - 2 = 10$. The correct choice is **(2)**.

7. The x will be eliminated after combining two equations when the coefficient of the x in one of the equations is the opposite of the coefficient of the x in the other equation. For choice (1), if both sides of the top equation are multiplied by 2, it would become $24x - 6y = 42$. $+24$ is not the opposite of -2. For choice (2) if both sides of the second equation are multiplied by 6, it becomes $-12x + 36y = 12$. Since -12 is the opposite of $+12$, this is the best answer. The correct choice is (**2**).

8. Multiply both sides of the bottom equation by -2 to get $-8x - 6y = -12$. Add this to the top equation to eliminate the x and get $-8y = 16$. Divide both sides by -8 to get $y = -2$. Substitute -2 for y into one of the original equations. $8x - 2(-2) = 28$, $8x + 4 = 28$, $8x = 24$, $x = 3$. The correct choice is (**4**).

9. If x is the number of two-legged animals and y is the number of four-legged animals, the number of animals is $x + y$ and the number of legs is $2x + 4y$. The system, then, is $x + y = 50$ and $2x + 4y = 172$. The correct choice is (**4**).

10. If x is the number of dogs and y is the number of cats, the system of equations is

$$x + y = 30$$
$$100x + 50y = 1,900$$

To eliminate the x, multiply both sides of the top equation by -100 to get $-100x - 100y = -3,000$. Add this to the bottom equation to get $-50y = -1,100$ or $y = 22$. The correct choice is (**3**).

5. GRAPHS OF LINEAR EQUATIONS

5.1 GRAPHING THE SOLUTION SET OF A LINEAR EQUATION BY MAKING A TABLE OF VALUES

The equation $x + y = 10$ has an infinite number of ordered pairs that satisfy it. One way to organize the information before creating a graph is to make a table of values.

This is a chart with three ordered pairs satisfying the equation $x + y = 10$. For a linear equation, only two ordered pairs are needed, but it is wise to do an extra ordered pair in case one of your first two is incorrect.

x	y
2	8
3	7
9	1

Plot the ordered pair $(2, 8)$ on the coordinate plane by locating the point that is two units to the right of the y-axis and 8 units above the x-axis. One way to do this is to start at the origin point where the two axes intersect and move to the right two units from there and then up 8 units.

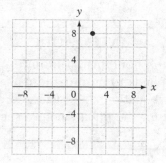

Do the same for the other two ordered pairs on the chart (3, 7) and (9, 1).

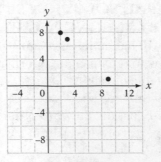

Draw a line through the three points. If the three points do not all lie on the same line, one of your ordered pairs is incorrect.

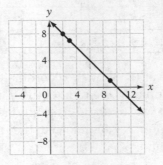

MATH FACTS

The line on a graph contains an infinite number of points. Each point corresponds to an ordered pair that is part of the solution set for the equation, and each ordered pair that is part of the solution set for the equation corresponds to a point on the line.

5.2 GRAPHING A LINEAR EQUATION USING THE INTERCEPT METHOD

When a linear equation is written in the form $ax + by = c$, the graph of the equation can be created quickly by finding the x-intercept and the y-intercept.

For the equation $2x - 3y = 12$:

- To find the y-intercept, substitute 0 for x and solve for y.

$$2(0) - 3y = 12$$
$$\frac{-3y}{-3} = \frac{12}{-3}$$
$$y = -4$$

The y-intercept is $(0, -4)$.

- To find the x-intercept, substitute 0 for y and solve for x.

$$2x - 3(0) = 12$$
$$2x = 12$$
$$x = 6$$

The x-intercept is $(6, 0)$.

- Plot $(0, -4)$ and $(6, 0)$ on a set of coordinate axes and draw the line that passes through both points.

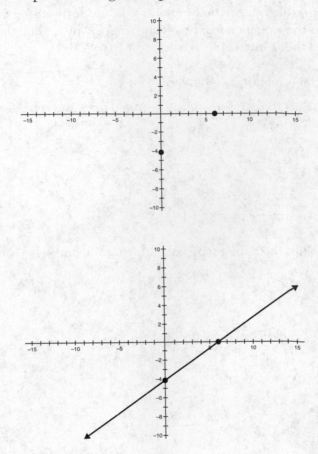

5.3 CALCULATING AND INTERPRETING SLOPE

The *slope* of a line is a number that measures how steep it is. A horizontal line has a slope of 0. A line with a positive slope goes up as it goes to the right. A line with a negative slope goes down as it goes to the right. The variable used for slope is the letter *m*.

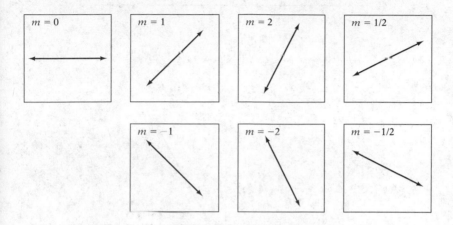

5.4 GRAPHING A LINEAR EQUATION IN SLOPE-INTERCEPT FORM

When the linear equation is in the form $y = mx + b$, it can be graphed very quickly. The y-intercept is $(0, b)$, and from that point to another point move one unit to the right and m units up.

For the equation $y = 2x - 3$, the y-intercept is $(0, -3)$. From that point, move 1 unit to the right and 2 units up. Draw the line through the two points $(0, -3)$ and $(1, -1)$.

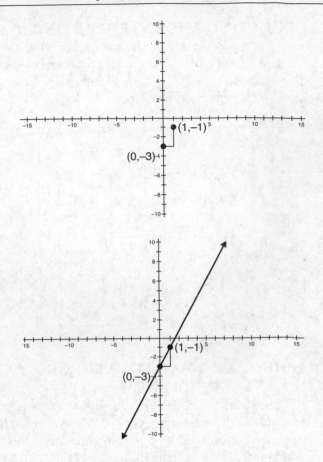

- If m is a fraction, n/d, it is more accurate to move d units to the right and n units up from the y-intercept.

- For the equation $y = \dfrac{2}{3}x + 2$, the y-intercept is $(0, 2)$. From that point move 3 to the right and 2 up to get to the point $(3, 4)$. Draw a line through these two points.

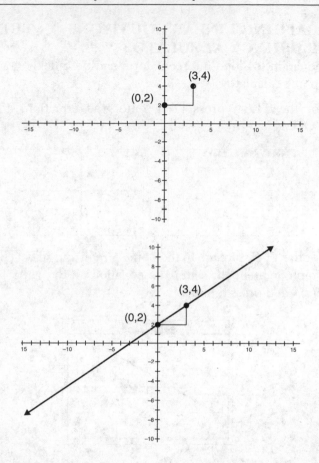

If the m value is negative, move down instead of up to get from the y-intercept to the next point.

5.5 GRAPHING LINEAR EQUATIONS ON THE GRAPHING CALCULATOR

An equation in slope-intercept form can be quickly graphed on the graphing calculator.

- For the TI-84, press [Y=], enter the equation, and press [ZOOM] and [6].

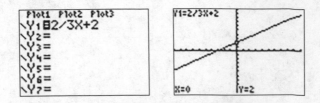

- For the TI-Nspire, go to the home screen and select [B] for the Graph Scratchpad. Enter the equation on the entry line after $f1(x)=$ and press [enter].

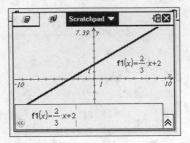

- If a linear equation with two variables is not originally in the slope-intercept form, $y = mx + b$, algebra can be used to rewrite it in slope-intercept form before graphing.

5.6 EQUATIONS OF VERTICAL AND HORIZONTAL LINES

The graph of the equation $y = k$ is a horizontal line through the point $(0, k)$. The graph of the equation $x = h$ is a vertical line through the point $(h, 0)$.

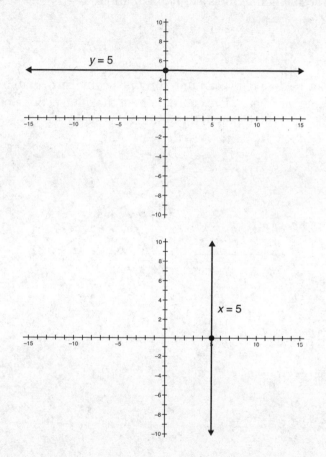

5.7 SOLVING SYSTEMS OF LINEAR EQUATIONS GRAPHICALLY

The solution to a system of equations is the set of coordinates of the point where the graphs of the two lines intersect.

- The solution to the system of equations

$$y = 2x - 1$$
$$y = -\frac{2}{3}x + 7$$

can be found by carefully producing the two graphs on the same set of axes. The coordinates of the intersection point $(3, 5)$ means the solution to the system of equations is $x = 3$, $y = 5$.

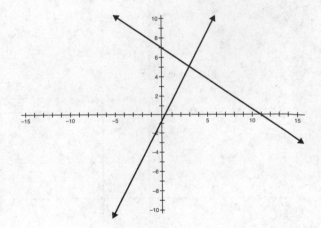

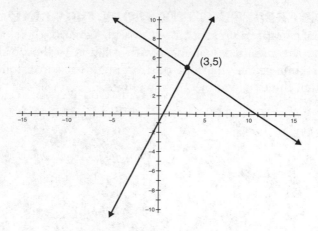

The graphing calculator can also determine the intersection point of two lines. On the TI-84, graph both lines and press [2ND], [TRACE], and [5] to find the intersection. On the TI-Nspire, graph both lines and press [menu], [6], and [4] to find the intersection.

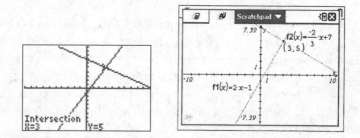

5.8 DETERMINING THE SLOPE OF A LINE

If two points on a line are (x_1, y_1) and (x_2, y_2), then the slope of the line can be determined by the formula

$$m = \frac{y_2 - y_1}{x_2 - x_1}$$

The slope of the line passing through $(3, 5)$ and $(7, 8)$ is

$$m = \frac{8-5}{7-3} = \frac{3}{4}$$

5.9 INTERPRETING THE SLOPE OF A LINE

When a graph represents a real-world scenario, the slope is the rate that the x value is changing with relation to the y value. In a distance-time graph, the slope of a line segment corresponds to the speed that the moving object is traveling.

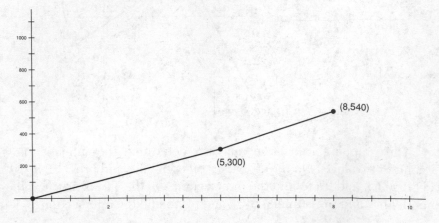

In the distance-time graph above, a car's time traveled is represented by the x-coordinate, and the distance it has covered at that time is represented by the y-coordinate. The slope of the segment connecting $(0, 0)$ and $(5, 300)$ is 60 so the car was traveling at a speed of 60 mph for the first 5 hours. The slope of the segment connecting $(5, 300)$ and $(8, 540)$ is 80 so the car was traveling at a speed of 80 mph for the last 3 hours.

5.10 FINDING THE EQUATION OF A LINE THROUGH TWO GIVEN POINTS

If two points are known, use the slope-intercept formula to find the slope of the line. Substitute the value you calculated for m and the x- and y-coordinates of either of the given two points into the equation $y = mx + b$ and solve for b. Substitute the values calculated for m and b into the equation $y = mx + b$.

- For the line through the two points $(3, 5)$ and $(9, 1)$, m is the slope of the line = $\dfrac{1-5}{9-3}$ = $-\dfrac{4}{6}$ = $-\dfrac{2}{3}$.

- Using the point $(3, 5)$, substitute $x = 3$, $y = 5$, $m = -\dfrac{2}{3}$ into the equation $y = mx + b$ and solve for b.

$$5 = -\frac{2}{3}(3) + b$$
$$5 = -2 + b$$
$$+2 = +2$$
$$7 = b$$

$m = -\dfrac{2}{3}$ and $b = 7$ so the equation is $y = -\dfrac{2}{3}x + 7$.

Practice Exercises

1. What is the x-intercept of the graph of the solution set of the equation $2x + 5y = 20$?
   (1) $(10, 0)$ (3) $(0, 10)$
   (2) $(4, 0)$ (4) $(0, 4)$

2. This is a graph of the solution set of which equation?

$$\frac{5y}{5} = \frac{-2x}{5} + \frac{20}{5}$$

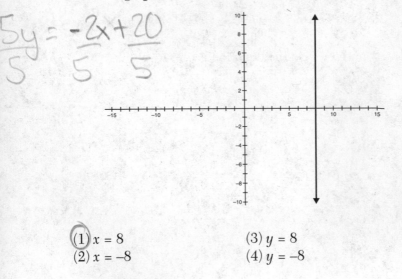

   (1) $x = 8$ (3) $y = 8$
   (2) $x = -8$ (4) $y = -8$

3. Below is the graph of the solution set of an equation. Based on this graph, which ordered pair does not seem to be part of the solution set of the equation $y = \dfrac{1}{3}x + 2$?

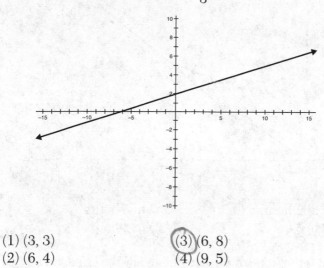

(1) $(3, 3)$ (3) $(6, 8)$

(2) $(6, 4)$ (4) $(9, 5)$

4. What is the slope of the line that passes through $(-2, 1)$ and $(8, 5)$?

(1) $\dfrac{2}{5}$ (3) $\dfrac{5}{2}$

(2) $-\dfrac{2}{5}$ (4) $-\dfrac{5}{2}$

5. Below is a distance-time graph for a bicycle trip. During which time interval is the cyclist going the fastest?

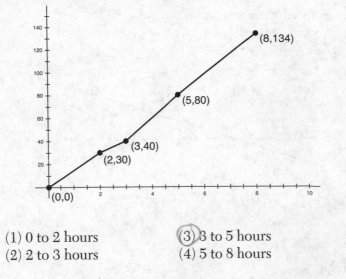

(1) 0 to 2 hours   (3) 3 to 5 hours
(2) 2 to 3 hours   (4) 5 to 8 hours

6. What is the slope of the line defined by the equation $y = -3x + 4$?
(1) 3   (3) 4
(2) –3   (4) –4

7. This is the graph of the solution set of which equation?

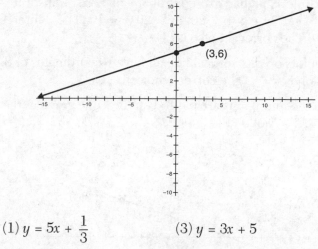

(1) $y = 5x + \dfrac{1}{3}$ (3) $y = 3x + 5$

(2) $y = 5x + 3$ (4) $y = \dfrac{1}{3}x + 5$

8. Find the equation of the line passing through the two points $(0, -7)$ and $(5, 8)$.

(1) $y = 3x + 7$ (3) $y = \dfrac{1}{3}x + 7$

(2) $y = 3x - 7$ (4) $y = \dfrac{1}{3}x - 7$

9. Find the equation of the line passing through the two points $(4, -2)$ and $(12, 4)$.

(1) $y = \dfrac{4}{3}x - 5$ (3) $y = \dfrac{3}{4}x - 5$

(2) $y = \dfrac{4}{3}x + 5$ (4) $y = \dfrac{3}{4}x + 5$

10. Find the equation of the line through the points $(3, 5)$ and $(3, 8)$.
(1) $x = -3$ (3) $y = -3$
(2) $x = 3$ (4) $y = 3$

Solutions

1. The x-intercept has a y-coordinate of zero. Substitute zero for y to get $2x + 5(0) = 20$, $2x = 20$, $x = 10$. The x-intercept is $(10, 0)$. The correct choice is **(1)**.

2. The points on this vertical line all have x-coordinates of 8. The equation is $x = 8$. The correct choice is **(1)**.

3. When the four points are plotted on the same graph as the line, only $(6, 8)$ is not on the line.

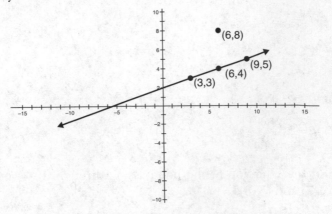

The correct choice is **(3)**.

4. Using the slope formula $m = \dfrac{y_2 - y_1}{x_2 - x_1}$, with the points $(-2, 1)$ and $(8, 5)$ becomes $m = \dfrac{5-1}{8-(-2)} = \dfrac{4}{10} = \dfrac{2}{5}$. The correct choice is **(1)**.

5. The interval between 0 and 2 has a slope of $30/2 = 15$. The interval between 2 and 3 has a slope of $10/1 = 10$. The interval between 3 and 5 has a slope of $40/2 = 20$. The interval between 5 and 8 has a slope of $54/3 = 18$. Since the slope represents the speed the bicycle is going, the interval between 3 and 5 hours is the fastest. It can also be seen from the graph that the interval between 3 and 5 hours looks the steepest. The correct choice is **(3)**.

6. When the equation for a line is in $y = mx + b$ form, the m is the slope. For this equation, the coefficient of the x is -3 so the slope of the line is -3. The correct choice is **(2)**.

7. According to the graph, the y-intercept is $(0, 5)$. The slope of the line through $(0, 5)$ and $(3, 6)$ is $m = \dfrac{6-5}{3-0} = \dfrac{1}{3}$. In $y = mx + b$ form, the m is the slope and the b is the y-coordinate of the y-intercept so the equation is $y = \dfrac{1}{3}x + 5$. The correct choice is **(4)**.

8. The slope of the line is $m = \dfrac{8-(-7)}{5-0} = \dfrac{15}{5} = 3$. Choose one of the points for the x and the y value and 3 for the m value. Using the point $(5, 8)$ the equation $y = mx + b$ becomes $8 = 3(5) + b$, $8 = 15 + b$, $b = -7$. The equation is $y = 3x - 7$. The correct choice is **(2)**.

9. The slope of the line is $m = \dfrac{4-(-2)}{12-4} = \dfrac{6}{8} = \dfrac{3}{4}$. Pick one of the points for x and y and substitute $\dfrac{3}{4}$ for m into $y = mx + b$. $4 = \dfrac{3}{4} \cdot 12 + b$, $4 = 9 + b$, $b = -5$. The equation is $y = \dfrac{3}{4}x - 5$. The correct choice is **(3)**.

10. Since the x-coordinates are the same, the line is a vertical line. Vertical lines have equations $x = $ constant. Since the x-coordinates are both 3, the equation is $x = 3$. The correct choice is **(2)**.

6. GRAPHS OF QUADRATIC EQUATIONS

6.1 GRAPHING A QUADRATIC EQUATION WITH A CHART

A **quadratic equation** can be written in the form $y = ax^2 + bx + c$. The graph of a quadratic equation is always a **parabola**, which resembles the letter U. When the a value is positive, the parabola looks like a right-side-up U. When the a value is negative, the parabola looks like an upside-down U.

- The simplest way to create the graph of a quadratic equation is to choose at least five consecutive x values and create a chart.

For the equation $y = x^2 - 2x - 3$, the chart could look like this:

x	y
-2	$(-2)^2 - 2(-2) - 3 = 4 + 4 - 3 = 5$
-1	$(-1)^2 - 2(-1) - 3 = 1 + 2 - 3 = 0$
0	$(0)^2 - 2(0) - 3 = -3$
1	$(1)^2 \ 2(1) \ 3 = 1 - 2 - 3 - -4$
2	$(2)^2 - 2(2) - 3 = 4 - 4 - 3 = -3$

- The graph of these five ordered pairs looks like this:

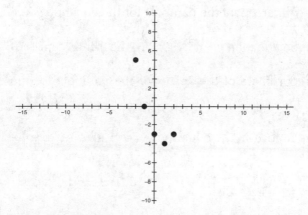

- *Connect* the points with a U-shaped parabola.

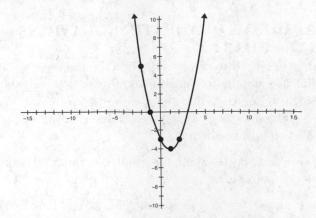

The low point of the parabola is called the **vertex**. In this example the vertex is $(1, -4)$. The two x-intercepts of the parabola are $(-1, 0)$ and $(3, 0)$. The y-intercept of this parabola is $(0, -3)$.

6.2 GRAPHING A PARABOLA BY FINDING THE VERTEX AND INTERCEPTS

The coordinates of the vertex, the y-intercept, and the x-intercepts can be calculated. These four points help produce an accurate graph of the parabola.

- The y-intercept of the parabola for the equation $y = ax^2 + bx + c$ is $(0, c)$.
- For the equation $y = x^2 - 2x - 3$, the y-intercept is $(0, -3)$.

The x-coordinate of the vertex of the parabola for the equation $y = ax^2 + bx + c$ is $x = -\dfrac{b}{2a}$. To get the y-coordinate, substitute the $-\dfrac{b}{2a}$ value in for x in the equation $y = ax^2 + bx + c$ to solve for y.

- For the equation $y = x^2 - 2x - 3$, the x-coordinate of the vertex is $x = -\dfrac{-2}{2(1)} = \dfrac{2}{2} = 1$. The y-coordinate is $y = 1^2 - 2(1) - 3 = 1 - 2 - 3 = -4$. So the vertex is $(1, -4)$.

The x-intercepts of the parabola for the equation $y = ax^2 + bx + c$ can be found by solving the quadratic equation $0 = ax^2 + bx + c$.
For the equation $y = x^2 - 2x - 3$, this becomes

$$0 = x^2 - 2x - 3$$

This quadratic equation can be solved by factoring.

$$0 = (x + 1)(x - 3)$$
$$x + 1 = 0 \text{ or } x - 3 = 0$$
$$-1 = -1 \qquad +3 = +3$$
$$x = -1 \text{ or } \qquad x = 3$$

The x-intercepts are $(-1, 0)$ and $(3, 0)$.

These four points help make an accurate sketch of the parabola.

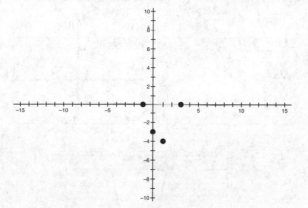

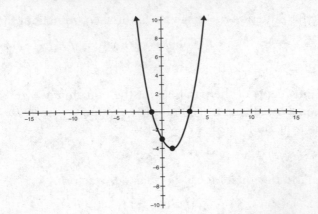

6.3 GRAPHING QUADRATIC EQUATIONS
ON THE GRAPHING CALCULATOR

The graphing calculator can easily graph quadratic equations. On the TI-84 press [Y=] and enter the equation after "Y1=", and press [ZOOM] and [6]. On the TI-Nspire, from the home screen press [B] for the Graph Scratchpad. Then enter the equation on the entry line and press [enter].

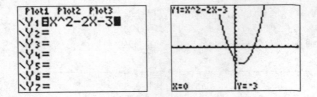

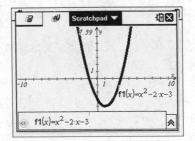

The graphing calculator can also determine the vertex and the intercepts using the min/max feature and the zeros feature.

- For the TI-84, press [2ND], [TRACE], and [3].

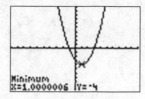

- For the TI-Nspire, press [menu], [6], and [2] to find the minimum point.

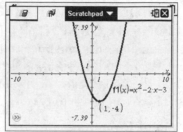

6.4 USING THE GRAPHING CALCULATOR TO SOLVE QUADRATIC EQUATIONS

The solutions to the equation $ax^2 + bx + c = 0$ are also the x-coordinates of the x-intercepts of the parabola defined by $y = ax^2 + bx + c$.

- To solve the equation $x^2 - 2x - 3 = 0$ with the graphing calculator, graph $y = x^2 - 2x - 3$ and then use the zeros feature to find the x-intercepts.
- For the TI-84, enter the equation $Y1 = x^2 - 2x - 3$ and [ZOOM] and [6] to graph. Then press [2ND], [TRACE], and [2] to find each zero.

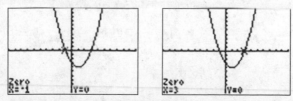

- For the TI-Nspire, enter the equation $f1(x) = x^2 - 6x + 8$ to graph it. Then press [menu], [6], and [1] to find each zero.

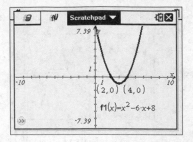

6.5 SOLVING A LINEAR-QUADRATIC SYSTEM OF EQUATIONS BY GRAPHING

When a system of equations has one linear equation and one quadratic equation, one way to find the solution is to graph the line and the parabola and find the coordinates of the intersection point, or intersection points. There can be up to two solutions.

- **For the System of Equations**

$$y = -2x + 1$$
$$y = x^2 - 2x - 3$$

- Create the graph for both equations either by hand or with the graphing calculator. The coordinates of the two intersection points are the two solutions to the system of equations. For this system the two solutions are $x = -2, y = 5$ and $x - 2, y = -3$.

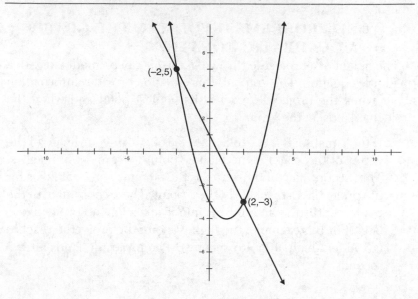

This can also be done on the graphing calculator.

- For the TI-84, graph the two equations and press [2ND], [TRACE], and [5] for each intersection point.

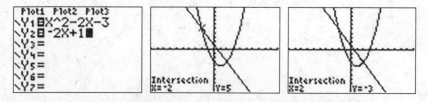

- For the TI-Nspire, graph the two equations and press [menu], [6], and [4] to find the intersection points.

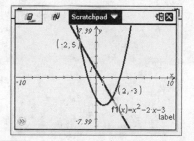

6.6 WORD PROBLEMS INVOLVING THE GRAPH OF A QUADRATIC EQUATION

The height of a projectile thrown in the air can be modeled with a quadratic equation. The graph of this equation provides information about when the projectile reaches its highest point and when the projectile lands on the ground.

- If the equation is $h = -16t^2 + 48t + 160$, the y-coordinate of the vertex of the graph is the highest point the projectile reaches. The x-coordinate of the vertex is the amount of time it takes for the projectile to reach its highest point. The x-coordinate of the x-intercept is the amount of time it takes for the projectile to land. For this example, the highest point the projectile reaches is 196 feet high after 1.5 seconds. The projectile lands after 5 seconds.

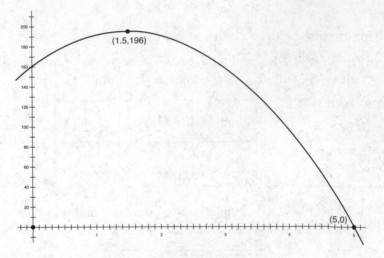

Practice Exercises

1. Which is a point on the graph of the solution set of $y = x^2 + 5x - 2$?
   (1) (3, 19) (3) (3, 21)
   (2) (3, 20) (4) (3, 22)

2. What are the coordinates of the vertex of the parabola defined by the equation $y = x^2 - 4x - 1$?
   (1) (-2, 5) (3) (-2, -5)
   (2) (2, 5) (4) (2, -5)

3. $x = -4$ is the x-coordinate of the vertex for the parabola defined by which equation?
   (1) $y = x^2 + 8x + 3$
   (2) $y = x^2 - 8x + 3$
   (3) $y = x^2 + 4x + 3$
   (4) $y = x^2 - 4x + 3$

4. What could be the equation that determines this parabola?

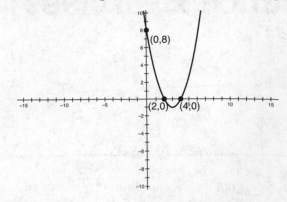

(1) $y = x^2 - 6x - 8$ (3) $y = x^2 + 6x - 8$

(2) $y = x^2 - 6x + 8$ (4) $y = x^2 + 6x + 8$

5. Which is the graph of $y = -x^2 + 2x + 3$?

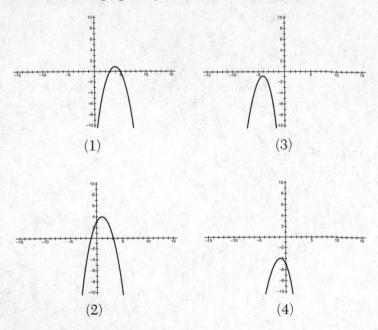

(1)   (3)

(2)   (4)

6. Which ordered pair is a solution to the system?

$$y = x^2$$
$$y = x + 2$$

(1) (1, 4)   (3) (3, 4)
(2) (2, 4)   (4) (4, 4)

7. Solve this system of equations using algebra.

$$y = x^2$$
$$y = 2x + 3$$

(1) (–1, 1) and (9, 3)   (3) (–1, 1) and (–3, 9)
(2) (–1, 1) and (3, 9)   (4) (1, –1) and (9, 3)

8. Which system of equations could be used to solve this graph?

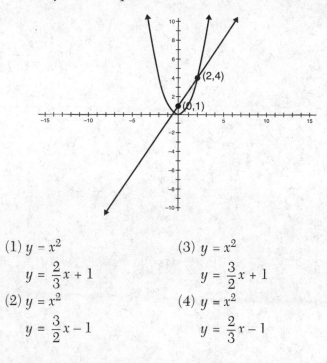

(1) $y = x^2$

$\quad y = \dfrac{2}{3}x + 1$

(2) $y = x^2$

$\quad y = \dfrac{3}{2}x - 1$

(3) $y = x^2$

$\quad y = \dfrac{3}{2}x + 1$

(4) $y = x^2$

$\quad y = \dfrac{2}{3}x - 1$

9. The x-intercepts of the parabola defined by which equation are the solutions to the equation $x^2 + 5x = 15$?
   (1) $y = x^2 + 5x + 15$ (3) $y = x^2 + 5x - 15$
   (2) $y = x^2 + 5x$ (4) $y = x^2 - 5x - 15$

10. Solve for all values of x, rounded to the nearest hundredth, $x^2 + 10x + 23 = 0$.
    (1) –3.41, –6.58   (3) –3.31, –6.64
    (2) –3.59, –6.41   (4) –3.62, –6.18

Solutions

1. For each choice, substitute the x-coordinate in for x and the y-coordinate in for y to see which makes the equation true. 22 does equal $3^2 + 5(3) - 2$ so the solution is $(3, 22)$. The correct choice is **(4)**.

2. The x-coordinate of the vertex of a parabola $y = ax^2 + bx + c$ is $x = \dfrac{-b}{2a}$. Since a is 1 and b is -4, the x-coordinate of the vertex is $x = -\dfrac{(-4)}{2(1)} = 2$. To get the y-coordinate of the vertex, substitute 2 for x into the equation and solve for y. $y = 2^2 - 4(2) - 1 = 4 - 8 - 1 = -5$. The vertex is $(2, -5)$. The correct choice is **(4)**.

3. Graph each choice on the graphing calculator and use the minimum feature to find the vertex of each. The vertex of the parabola $y = x^2 + 8x + 3$ is $(-4, -15)$, which has an x-coordinate of -4. The correct choice is **(1)**.

4. Since the x-intercepts are $(2, 0)$ and $(4, 0)$, the equation for the parabola is $y = a(x - 2)(x - 4) = a(x^2 - 6x + 8)$. Since the y-intercept is $+8$, $a = 1$ and the equation is $y = x^2 - 6x + 8$. The correct choice is **(2)**.

5. The x-coordinate of the vertex must be $x = -\dfrac{2}{2(1)} = 1$. The correct choice is **(2)**.

6. Substitute x^2 for y in the bottom equation to get $x^2 = x + 2$, $x^2 - x - 2 = 0$, $(x - 2)(x + 1) = 0$, $x = 2$ or $x = -1$. The x-coordinates of the intersection points are 2 and -1. This can also be done with the graphing calculator. The answer is $(2, 4)$. The correct choice is **(2)**.

7. Substitute x^2 for y in the bottom equation to get $x^2 = 2x + 3$, $x^2 - 2x - 3 = 0$, $(x - 3)(x + 1) = 0$. $x = 3$ and $x = -1$ are the x-coordinates of the two intersection points. The correct choice is **(2)**.

8. The parabola with vertex $(0, 0)$ passing through $(2, 4)$ is $y = x^2$. The line through $(0, 1)$ and $(2, 4)$ is $y = (3/2)x + 1$. The correct choice is **(3)**.

9. This equation can be rewritten as $x^2 + 5x - 15 = 0$. Equations in this form can be solved by finding the x-intercepts of the parabola $y = x^2 + 5x - 15$. The correct choice is **(3)**.

10. Using the graphing calculator, graph $y = x^2 + 10x + 23$ and use the "zero" feature to get $x = 3.59$ or $x = -6.41$. The correct choice is **(2)**.

7. LINEAR INEQUALITIES

7.1 ONE-VARIABLE LINEAR INEQUALITIES

A linear inequality is like a linear equation, but instead of an = sign there is either a >, <, ≥, or ≤ sign. Solving a one-variable linear inequality is almost the same as solving a linear equality. The only difference is that when multiplying or dividing both sides by a negative number to eliminate the coefficient, the direction of the inequality sign must be reversed.

$$-2x + 3 < 11$$

- Eliminate the constant 3 by subtracting it from both sides of the inequality.

$$-2x + 3 < 11$$
$$-3 = -3$$
$$-2x < 8$$

- *Divide* both sides by –2 to isolate the x. Because you are dividing by a negative, the direction of the inequality sign must be reversed to keep the equation true. If the –2 were a +2, the direction of the inequality sign would not need to be reversed.

$$\frac{-2x}{-2} < \frac{8}{-2}$$
$$x > -4$$

The solution is $x > -4$.

7.2 GRAPHING TWO-VARIABLE INEQUALITIES

$y \leq 2x - 5$ is a two-variable inequality. To graph the two-variable inequality, first graph the line $y = 2x - 5$.

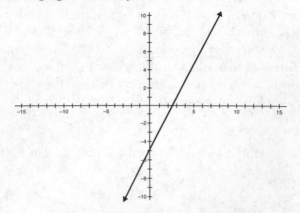

One side of the line needs to be shaded in. To determine which side of the line to shade, substitute the ordered pair $(0, 0)$ into the inequality to test if it yields a true inequality.

$$0 < 2(0) - 5$$
$$0 < 0 - 5$$
$$0 < -5 \text{ is not true.}$$

Since $(0, 0)$ does not satisfy the inequality, shade the side of the line that does not contain $(0, 0)$.

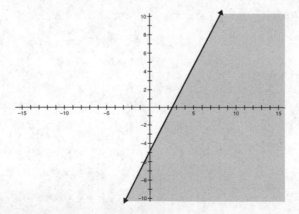

- If the inequality sign is a < or a >, the line must be a dotted line. If it is a ≤ or a ≥, it must be a solid line. If $(0, 0)$ is on the line, an ordered pair that is not on the line must be used to test which side to shade.
- For the graph of $y > 3x$, graph the line $y = 3x$ and make it a dotted line.

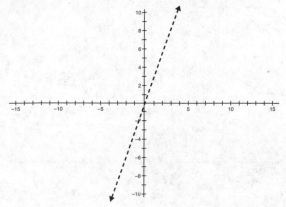

Since $(0, 0)$ is on the line, test a point that is not on the line, like $(2, 0)$.

$$0 > 3(2)$$

Since $0 > 6$ is not true, shade the side that does not contain $(2, 0)$.

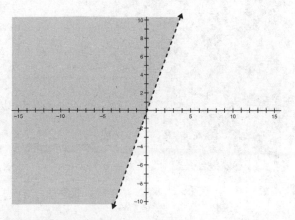

7.3 GRAPHING SYSTEMS OF LINEAR INEQUALITIES

To graph the solution set to a system of linear inequalities, graph both inequalities on the same graph and locate the portion of the graph that is shaded twice.

$$y > 2x - 5$$

$$y < -\frac{2}{3}x + 2$$

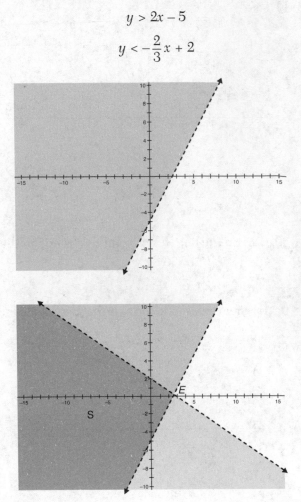

The portion of the graph that has an S is the solution set. Any point in that region will satisfy the system of inequalities. (−3, 1) is an example of a point that is in this double-shaded region and will satisfy both inequalities.

7.4 GRAPHING INEQUALITIES ON THE GRAPHING CALCULATOR

The TI-84 and the TI-Nspire can graph linear inequalities.

- On the TI-84, the shading can be set by moving the cursor to the icon to the left of the Y= and pressing [ENTER] until the icon shows the proper shading.

- On the TI-Nspire, delete the "=" from the equation in the entry line and choose the symbol to replace it with.

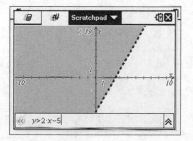

The graphing calculator can also graph systems of linear inequalities by graphing both inequalities on the same set of coordinate axes.

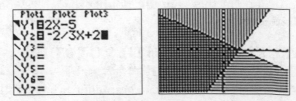

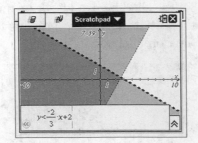

Practice Exercises

1. What is the solution set for $3x < -18$?
   (1) $x < -6$
   (2) $x > -6$
   (3) $x \geq -6$
   (4) $x \leq -6$

2. What is the solution set for $-4x \geq 20$?
   (1) $x \geq -5$
   (2) $x > -5$
   (3) $x \leq -5$
   (4) $x < -5$

3. What is the smallest integer that satisfies the equation $-6x < -18$?
   (1) 3
   (2) 4
   (3) 5
   (4) 6

4. Which is the graph of $y < x + 4$?

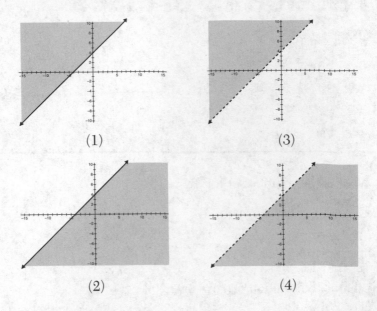

(1)   (3)

(2)   (4)

5. Below is the graph of $x + y \leq 8$. Which is a point in the solution set?

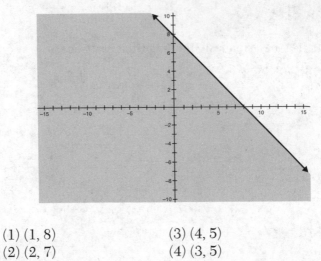

(1) (1, 8)   (3) (4, 5)
(2) (2, 7)   (4) (3, 5)

6. All of these ordered pairs are part of the shaded region for the graph of $2x + y \le 12$ except

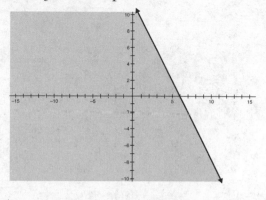

(1) $(2, 8)$ (3) $(5, 3)$

(2) $(3, 6)$ (4) $(4, 4)$

7. Which graph shows the solution to the system of inequalities?

$$y < 2x + 1$$

$$y > \frac{1}{3}x + 4$$

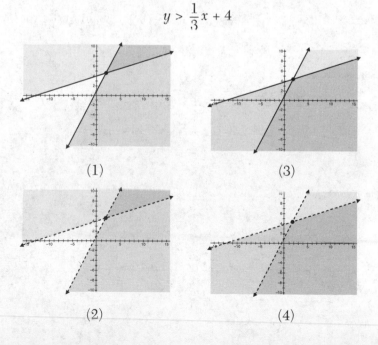

8. Which system of inequalities does the following graph show the solution for?

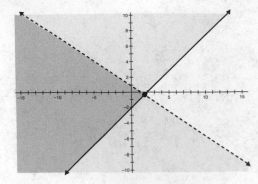

(1) $y \geq x - 2$

$\quad y < -\dfrac{2}{3}x + 1$

(2) $y \leq x - 2$

$\quad y > -\dfrac{2}{3}x + 1$

(3) $y < x - 2$

$\quad y \geq -\dfrac{2}{3}x + 1$

(4) $y > x - 2$

$\quad y \leq -\dfrac{2}{3}x + 1$

9. Which graph has the solution set shaded in for the following system of inequalities?

$$y \leq -x + 6$$

$$y \geq \frac{1}{2}x - 1$$

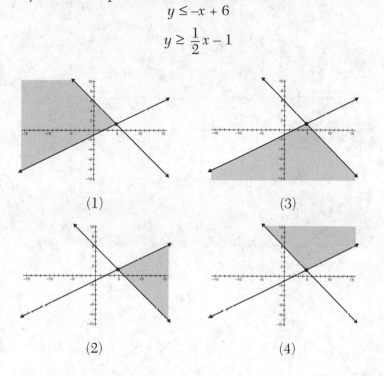

(1)   (3)

(2)   (4)

10. Which is the graph of the following system of inequalities?

$$y \geq 0$$
$$x \leq 0$$

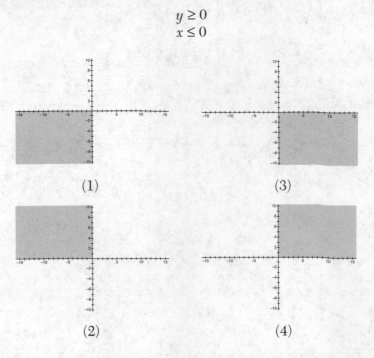

(1)

(3)

(2)

(4)

Solutions

1. Divide both sides of the inequality by 3 to get $x < -6$. The correct choice is **(1)**.

2. Divide both sides of the inequality by -4. Switch the direction of the inequality sign because you divided by a negative to get $x \leq -5$. The correct choice is **(3)**.

3. Divide both sides of the inequality by -6 and switch the direction of the inequality sign to get $x > 3$. The smallest integer that satisfies the equation is 4. The answer is not 3 since it is not true that $3 > 3$. The correct choice is **(2)**.

4. Because it is a $<$ and not a $\leq$ sign, the line must be dotted. Test to see if $(0, 0)$ is in the solution set by checking if $0 < 0 + 4$ is true. Since it is true, the side of the line containing $(0, 0)$ must be shaded. The correct choice is **(4)**.

5. When all four choices are plotted on the graph, three of them are not in the shaded region. The point $(3, 5)$ is on the line, but since there is a $\leq$ sign, the line is part of the solution set. The correct choice is **(4)**.

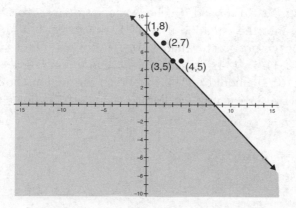

6. Three of the points are on the line, which is part of the solution set because it is a ≤ sign. The point (5, 3) is not in the shaded area. The correct choice is **(3)**.

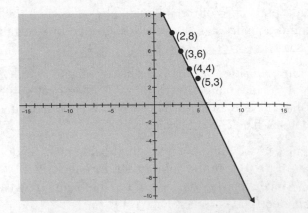

7. Both of the lines have to be dotted because of the inequality signs < and >. This eliminates choices 1 and 3. Both choices 2 and 4 have the region below the line $y = 2x + 1$ shaded. The difference between choices 2 and 4 is that choice 2 has the region above $y = (1/3)x + 4$ shaded, and choice 4 has the region below $y = (1/3)x + 4$ shaded. To check which side is correct, substitute (0,0) into the equation $y > (1/3)x + 4$ to get $0 > (1/3)(0) + 4$. Since 0 is not greater than 4, (0, 0) is not part of the solution set to $y > (1/3)x + 4$. The side of the line $y = (1/3)x + 4$ that contains (0, 0) should not be shaded so the region above the line $y = (1/3)x + 4$ should be shaded. The correct choice is **(2)**.

8. Since the line $y = x - 2$ is solid, choices 3 and 4 can be elimi-
    nated. Substituting $(0, 0)$ into both inequalities in choice 1 makes
    them both true while substituting $(0, 0)$ into both inequalities
    in choice 2 makes them both false. The correct choice is **(1)**.

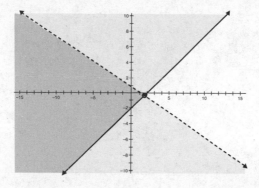

9. To check choice 1, pick a point in the shaded region for that
    choice and check to see if it satisfies both inequalities. Since $(0, 0)$
    is a point in the shaded region, check to see if $0 \le -0 + 6$ and
    $0 \ge (1/2) \cdot 0 - 1$. $(0, 0)$ does make both inequalities true. The cor-
    rect choice is **(1)**.

10. The points on or above the x-axis make $y \ge 0$ true. The points
    on or to the left of the y-axis make $x \le 0$ true. The points that
    satisfy both inequalities must be above the x-axis and to the left
    of the y-axis. The correct choice is **(2)**.

8. EXPONENTIAL EQUATIONS

8.1 EVALUATING EXPONENTIAL EXPRESSIONS

An **exponential equation** is one where one of the variables is an exponent. The equation $y = 3 \cdot 2^x$ is a two-variable exponential equation. Exponential equations can be written in the form $y = a \cdot b^x$.

- When evaluating an *exponential expression*, raise the base to the exponent before multiplying. For example, to evaluate $y = 3 \cdot 2^x$ when $x = 4$ it becomes

$$y = 3 \cdot 2^4$$
$$y = 3 \cdot 16 \ (\text{NOT } y = 6^4)$$
$$y = 48$$

8.2 EXPONENTIAL GROWTH VS. EXPONENTIAL DECAY

In the exponential equation $y = a \cdot b^x$, the b is called the **base**. When b is greater than 1, the equation is an example of *exponential growth* since $a \cdot b^x$ grows as x grows. When b is between 0 and 1 the equation is an example of *exponential decay* since $a \cdot b^x$ gets smaller as x grows.

$y = 3 \cdot 2^x$ is an example of exponential growth since $b = 2$, which is greater than 1.

$y = 3 \cdot (0.7)^x$ is an example of exponential decay since $b = 0.7$, which is between 0 and 1.

8.3 GRAPHS OF EXPONENTIAL EQUATIONS

A table of values for a two-variable exponential equation, like $y = 3 \cdot 2^x$, looks like this:

x	y
-3	$3 \cdot 2^{-3} = 3 \cdot \left(\dfrac{1}{2^3}\right) = 3 \cdot \dfrac{1}{8} = \dfrac{3}{8}$
-2	$3 \cdot 2^{-2} = 3 \cdot \left(\dfrac{1}{2^2}\right) = 3 \cdot \dfrac{1}{4} = \dfrac{3}{4}$
-1	$3 \cdot 2^{-1} = 3 \cdot \left(\dfrac{1}{2^1}\right) = 3 \cdot \dfrac{1}{2} = \dfrac{3}{2}$
0	$3 \cdot 2^0 = 3 \cdot 1 = 3$
1	$3 \cdot 2^1 = 3 \cdot 2 = 6$
2	$3 \cdot 2^2 = 3 \cdot 4 = 12$
3	$3 \cdot 2^3 = 3 \cdot 8 = 24$

Remember that raising a number to a negative power, like a^{-x}, is equivalent to $\dfrac{1}{a^x}$. For example, $2^{-3} = \left(\dfrac{1}{2^3}\right) = \dfrac{1}{8}$.

The graph for this equation has a shape like this:

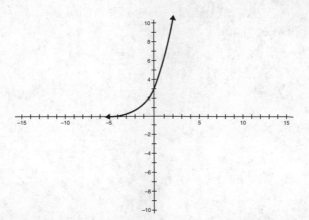

- For exponential decay, like $y = 3 \cdot \left(\dfrac{1}{2}\right)^x$, the graph has a shape like this:

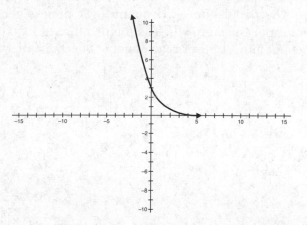

Exponential equations can also be graphed on the graphing calculator.

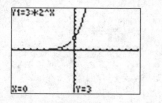

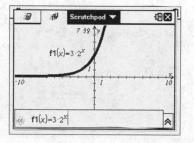

8.4 REAL-WORLD SCENARIOS INVOLVING EXPONENTIAL EQUATIONS

Many real-world scenarios can be modeled with exponential equations. The population of a country over time is generally an example of exponential growth. The temperature of food over time after being put into a freezer is an example of exponential decay.

Practice Exercises

1. If $x = 2$ and $y = 3^x$, solve for y.
   (1) 8
   (2) 9
   (3) 10
   (4) 11

2. If $x = 3$ and $y = 2 \cdot 3^x$, solve for y.
   (1) 216
   (2) 27
   (3) 54
   (4) 18

3. Which ordered pair is in the solution set of $y = 5 \cdot 2^x$?
   (1) (0, 0)
   (2) (2, 25)
   (3) (3, 40)
   (4) (3, 100)

4. Which is the graph of $y = 2^x$?

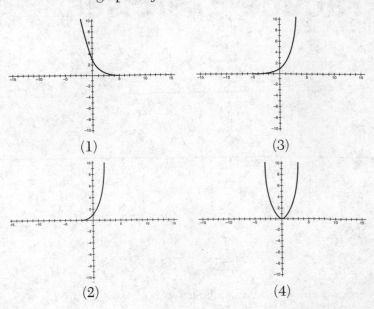

$$(1) \qquad\qquad\qquad (3)$$

$$(2) \qquad\qquad\qquad (4)$$

5. Below is the graph of which equation?

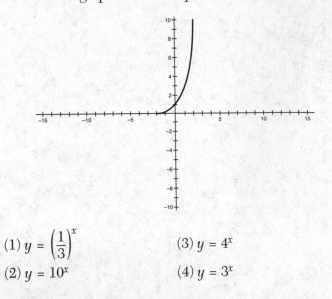

(1) $y = \left(\dfrac{1}{3}\right)^x$ (3) $y = 4^x$

(2) $y = 10^x$ (4) $y = 3^x$

6. In what interval is the graph of $y = 1.5^x$ increasing?
(1) Always
(2) Never
(3) When $x \geq 0$
(4) When $x \leq 0$

7. Below is the graph of $y = b^x$. What is true about the value of b?

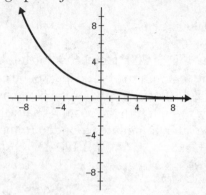

(1) b must be greater than 1
(2) b must be less than 1
(3) b must be less than 0
(4) b must be less than −1

8. What type of equation has a graph like the one below?

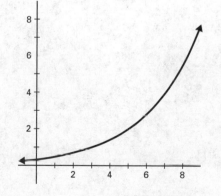

(1) Linear
(2) Exponential
(3) Quadratic
(4) None of the above

9. The population of a country can be modeled with the equation $P = 250 \cdot 1.07^t$, where P is the population in millions and t is the number of years since 2010. According to this model, rounded to the nearest ten million, what will the population of this country be in 2019?

    (1) 450,000,000   (3) 470,000,000
    (2) 460,000,000   (4) 480,000,000

10. A cup of tea that is 200 degrees is put into a room that is 80 degrees. The temperature of the tea can be calculated with the formula $t = 200 \cdot 0.9^m + 80$, where m is the number of minutes since the tea was put into the room. What will the temperature of the tea be after 15 minutes rounded to the nearest degree?

    (1) 116 degrees   (3) 120 degrees
    (2) 118 degrees   (4) 121 degrees

Solutions

1. Substitute 2 for x and the equation becomes $y = 3^2 = 9$. The correct choice is **(2)**.

2. Substitute 3 for x and the equation becomes $y = 2 \cdot 3^3 = 2 \cdot 27 = 54$. The correct choice is **(3)**.

3. Substitute each ordered pair into the equation to see which makes it true. For the ordered pair $(3, 40)$, $5 \cdot 2^3 = 5 \cdot 8 = 40$. The correct choice is **(3)**.

4. The graph for $y = 2^x$ must pass through the points $(0, 1)$, $(1, 2)$, and $(2, 4)$. It also passes through $(-1, 1/2)$ and $(-2, 1/4)$. The correct choice is **(3)**.

5. Since this graph passes through $(0, 1)$, $(1, 3)$, and $(2, 9)$, it is the equation $y = 3^x$. The correct choice is **(4)**.

6. An exponential graph with a base greater than 1 is always increasing. The correct choice is **(1)**.

7. When an exponential graph is decreasing, the base is between 0 and 1. The correct choice is **(2)**.

8. This graph has the standard shape of an exponential curve based on an equation $y = b^x$ with $b > 1$. An exponential curve like this starts off relatively flat for x values close to zero and then increases rapidly. The correct choice is **(2)**.

9. Since 2019 is 9 years past 2010, substitute 9 for t to get $P = 250 \cdot 1.07^9 - 459.6$. In millions, this is approximately $460,000,000$. The correct choice is **(2)**.

10. Substitute 15 for m to get $t = 200 \cdot 0.9^{15} + 80 = 121.1782$ or 121.2, which rounds to 121. The correct choice is **(4)**.

9. CREATING AND INTERPRETING EQUATIONS

9.1 CREATING AND INTERPRETING LINEAR EQUATIONS

Many real-world situations can be modeled with a linear equation in the form $y = mx + b$. The m and the b have to be replaced with the appropriate values from the situation. In general, the b value is the starting value and the m value is the amount the total changes each time the x variable increases.

- If a carnival costs \$10 admission and \$3 for each ride, the 10 and the 3 can be used in a linear equation. Since the 10 is the starting amount, it would take the place of the b in the equation. Since 3 is the amount the total increases by for each new ride, it would take the place of m in the equation.

  The equation could be written $y = 3x + 10$, where y is the total and x is the number of rides. Instead of x and y, the total could be represented by the variable T, whereas the number of rides could be represented by the variable R to form the equation $T = 3R + 10$.

  When given an equation modeling a real-world situation, it is also possible to interpret what the values for m and b represent. The b represents the starting value, and the m represents the amount that the total changes for each increase in x.

- If the equation for the cost of a pizza with N toppings is $C = 2N + 12$, you could be asked to interpret what the 2 and the 12 represent. Since the 2 is in the place of the m in $y = mx + b$, it represents the cost of each topping. Since the 12 is in the place of the b in $y = mx + b$, it represents the cost of the pizza before any toppings are added.

9.2 CREATING AND INTERPRETING EXPONENTIAL EQUATIONS

An **exponential equation** is a good model for many real-world situations including population growth, compound interest, and liquid cooling. Exponential equations have the form $y = a \cdot (1 + r)^x$, where the a represents the starting value and the r represents the growth rate.

- If the population of a town is 10,000 people and the annual growth rate is 7%, then the equation that relates total population, P, to the number of years that have passed, T, is $P = 10,000 \cdot 1.07^T$. Since 10,000 is the starting value, it takes the place of the a and since the growth rate is .07, it takes the place of the r in the equation $y = a \cdot (1 + r)^x$.

- If an exponential equation that models a real-world situation is given, it is possible to interpret what the numbers in the equation represent.

- If in another town, the equation relating their population to the number of years that have passed is $P = 20,000 \cdot 1.09^T$, you could be asked to interpret what the numbers 20,000 and .09 represent. In this case, the 20,000 represents the starting population and the .09 represents the growth rate.

- When a ball is dropped, each bounce is 80% as high as the previous bounce. If the ball is dropped from a window 50 feet above the ground, the equation that relates the height of the bounce (H) to the number of bounces (B) will be $H = 50 \cdot 0.80^B$. Since the starting height is 50, it takes the place of the a, and since 0.80 can be expressed as $(1 - 0.20)$, the r value is -0.20, which is the growth rate. A -0.20 growth rate can also be called a *decay* rate of 0.20.

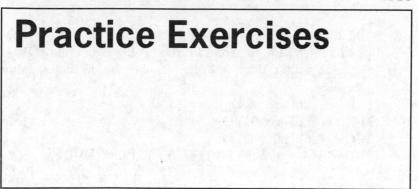

Practice Exercises

1. It costs \$10 to go to the movies and \$3 for each bag of popcorn. Which equation relates the total cost (C) to the number of bags of popcorn purchased (P)?
   (1) $C = 3P + 10$
   (2) $C = 10P + 3$
   (3) $P = 3C + 10$
   (4) $P = 10C + 3$

2. A tablet computer costs \$400 and \$2 for each app. Which equation relates the total cost (C) to the number of apps purchased (A)?
   (1) $A = 2 + 400C$
   (2) $A = 400 + 2C$
   (3) $C = 2 + 400A$
   (4) $C = 400 + 2A$

3. A cable TV plan costs \$80 a month plus \$10 extra for each premium channel. Which equation relates the monthly bill (B) to the number of premium channels ordered (C)?
   (1) $B = 80C + 10$
   (2) $B = 10C + 80$
   (3) $C = 80B + 10$
   (4) $C = 10B + 80$

4. Lydia wants to buy a DVD player and some DVDs. The equation that relates the total cost for the DVD player and N DVDs is $P = 20N + 200$. What does the number 200 in the equation represent?
   (1) The cost of the DVD player
   (2) The cost of each DVD
   (3) The cost of all N DVDs
   (4) The total cost of the DVD player and all N DVDs

5. Amelia buys an empty sticker album and some sticker sheets. The equation that relates the total cost for the empty sticker album and N sticker sheets is $P = 0.75N + 3.00$. What does the number 0.75 in the equation represent?
   (1) The cost of the empty sticker album
   (2) The cost of each sticker sheet
   (3) The cost of all N sticker sheets
   (4) The total cost of the empty sticker album and all N sticker sheets

6. There were 900 birds in a forest. Each year the bird population increases by 12%. Which equation relates the bird population (P) to the number of years that have passed?
   (1) $P = 900(1.12)^t$
   (2) $P = 900(0.12)^t$
   (3) $P = 900(0.88)^t$
   (4) $t = 900(1.12)^P$

7. A bouncing ball is dropped from 20 feet high. After each bounce, the height of the next bounce is 65% as high as the last bounce. Which equation relates the height of the bounce (H) to the number of bounces that have happened (N)?
   (1) $H = 20(0.35)^N$
   (2) $H = 20(1.65)^N$
   (3) $H = 20(0.65)^N$
   (4) $N = 20(0.65)^H$

8. The population (P) of a town after t years can be modeled with the equation $P = 20{,}000(1.07)^t$. What does the 20,000 represent?
   (1) The growth rate
   (2) The percent increase each year
   (3) The population after t years
   (4) The starting population of the town

9. After Allie takes some medicine, each hour the number of milligrams of medicine (M) remaining in her body after t minutes can be modeled with the equation $M = 200(1 - 0.27)^t$. Which number represents the decay rate?
   (1) 0.27
   (2) 200
   (3) 0.73
   (4) 146

10. Mason puts money into a bank that offers interest compounded annually. The formula relating the amount of money in the bank (A) to the number of years it has been in the bank (t) is $A = 800(1.2)^t$. What is the interest rate the bank offers?
    (1) 1.2%
    (2) 2%
    (3) 20%
    (4) 120%

Solutions

1. P bags of popcorn cost $\$3P$. The entrance price is $\$10$ so the total price is $3P + 10$. The correct choice is (**1**).

2. A apps cost $\$2A$. The computer costs $\$400$ so the total price is $2A + 400$ or $400 + 2A$. The correct choice is (**4**).

3. C premium channels cost $\$10C$. The monthly cost is $\$80$ so the total price is $10C + 80$. The correct choice is (**2**).

4. The 20 is the cost for each DVD and the 200 is the price of the DVD player. In general, the constant represents the fixed cost. The correct choice is (**1**).

5. The 3 is the part that does not depend on the value of N. The $0.75N$ increases by 0.75 each time N increases by 1 so the 0.75 represents the cost of each sticker sheet. The correct choice is (**2**).

6. In an exponential equation of the form $P = a \cdot (1 + r)^t$, the r value is the percent increase (or decrease if r is negative), and the a is the initial value. When $r = 0.12$ and $a = 900$, the equation is $P = 900(1.12)^t$. The correct choice is (**1**).

7. The height of each bounce is 0.65 multiplied by the height of the previous bounce. The initial height is 20 feet. After one bounce, the second is $20 \cdot 0.65$. After two bounces it has a height of $20 \cdot 0.65 \cdot 0.65$. In general, after N bounces, the height will be $H = 20 \cdot 0.65^N$. The correct choice is (**3**).

8. In an equation of the form $P = a \cdot (1 + r)^t$, the r is the percent increase each year and the a is the initial value at $t = 0$. The 20,000 then represents the starting population of the town. The correct choice is (**4**).

9. In an equation of the form $M = a \cdot (1 - r)^t$, the r is the decay rate. Since the $(1 - 0.27)$ is being raised to the t power in this equation, the decay rate is 0.27. The correct choice is (**1**).

10. The interest rate is the percent increase each year. The percent increase in an equation of the form $A = P \cdot (1 + r)^t$ is the r variable. Since $1.2 = 1 + 0.2$, the r value is 0.2, which is 20%. The correct choice is **(3)**.

10. FUNCTIONS

10.1 DIFFERENT REPRESENTATIONS OF FUNCTIONS

A **function** is like a machine that takes a number as an input and outputs a number. Functions are often named with lowercase letters like f and g.

- If a function f takes the number 2 as an input and outputs the number 7, we say $f(2) = 7$. The number in the parentheses is the number that is input into the function. The number after the equals sign is the number that is output from the function.

A function can be represented in several different ways. Here are the most common ways:

1. As a list of ordered pairs

- If the function f is defined as $f = \{(1, 4), (2, 7), (3, 10), (4, 13), (5, 16)\}$, the numbers in the parentheses represent an input value and an output value for the function. In this example, the point $(1,4)$ in the definition means that if 1 is put into the function, 4 is output from the function, or $f(1) = 4$. Likewise, $f(2) = 7, f(3) = 10, f(4) = 13,$ and $f(5) = 16$.

2. As an equation

- A function can be defined by an equation that allows you to calculate the output value for a given input value. An example is the definition $f(x) = 3x + 1$. With this definition, it is possible to calculate the output value for any input value. For example, to calculate $f(10)$, substitute the number 10 into the equation $f(x) = 3x + 1$ to become $f(10) = 3(10) + 1 = 30 + 1 = 31$ so $f(10) = 31$.

3. As a graph

- If the function is defined as a graph, determine the value of $f(2)$ by finding a point on the graph that has an x-coordinate of 2. The y-coordinate of that point is the value of $f(2)$. Since the point with x-coordinate of 2 is $(2, 7)$ the value of $f(2) = 7$.

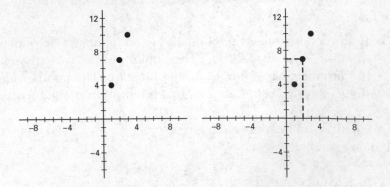

10.2 DOMAIN AND RANGE OF FUNCTIONS

The possible input values of a function are called the *domain* of the function. The possible output values are called the *range* of the function.

For the function $f = \{(1, 4), (2, 7), (3, 10), (4, 13), (5, 16)\}$, the domain is the set of input values $\{1, 2, 3, 4, 5\}$. The range is the set of output values $\{4, 7, 10, 13, 16\}$.

- When a function is described as a graph, the domain is the set of x-coordinates of all the points on the graph, and the range is the set of y-coordinates of all the points on the graph.

For this graph, the domain is {1, 2, 3, 4}, and the range is {3, 5, 8}.

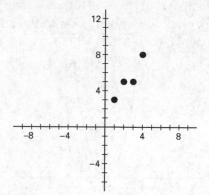

For this graph, the domain is $2 \leq x \leq 5$, and the range is $3 \leq y \leq 7$.

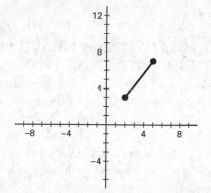

In a real-world situation, the domain is often a special subset of the real numbers. If the function has as its input the number of cars a salesman sells in a month, the domain would be the set of non-negative integers {0, 1, 2, ...} since you cannot sell fractions of a car or negative cars.

10.3 GRAPHING FUNCTIONS

The graph of the function $f(x) = x^2$ is the same as the graph $y = x^2$. All the methods for graphing by hand or with the graphing calculator described earlier can be used for functions.

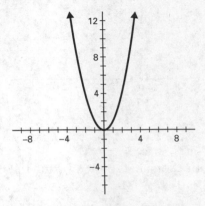

10.4 HOW TO TELL WHEN A GRAPH OR A LIST CANNOT DEFINE A FUNCTION

In a function, each time a number from the domain is put into the function, the same value is output from the function. So if $f(2) = 7$, the number 7 will always be output from the function whenever 2 is put into the function.

- When a function is defined as a list of ordered pairs, then there will never be two ordered pairs with the same x-coordinate, but different y-coordinates.

$f = \{(1, 4), (2, 7), (2, 10), (3, 13)\}$ is not the definition of a function since $f(2)$ can be 7 or 10.

$g = \{(1, 4), (2, 7), (3, 7), (4, 13)\}$ is the definition of a function. The fact that $g(2) = 7$ and $g(3) = 7$ does not contradict the definition of a function. As long as there are no repeats of x-coordinates, there can be repeats of y-coordinates.

A graph will not be the graph of a function if there are two points that have the same x-coordinate, but different y-coordinates. Graphs that cannot be functions fail the **vertical line test**, which means that at least one vertical line will pass through two or more points. All the points it passes through will have the same *x*-coordinates but different *y*-coordinates.

These graphs cannot be graphs of functions since they fail the vertical line test at at least one location.

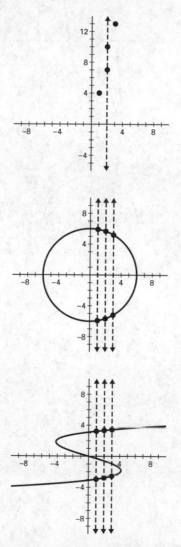

These graphs can be graphs of functions since they pass the vertical line test at all locations.

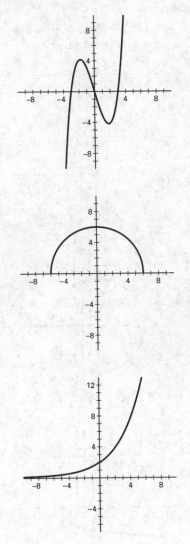

10.5 GRAPHING TRANSFORMED FUNCTIONS

- If the graph of $y = f(x)$ is already known, the graph of $y = f(x + a)$, $y = f(x - a)$, $y = f(x) + a$, and $y = f(x) - a$ can be easily graphed by knowing the four basic transformations.

- If the graph of $y = f(x)$ looks like this:

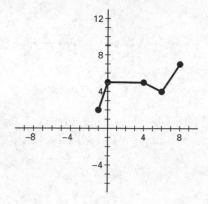

then the graph of

1. $f(x) + a$ will be the graph of $f(x)$ with every point shifted *up* by a units. The graph of $f(x) + 2$ looks like this:

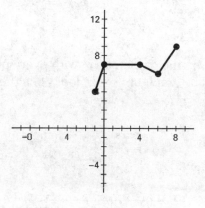

2. $f(x) - a$ will be the graph of $f(x)$ with every point shifted *down* by a units. The graph of $f(x) - 2$ looks like this:

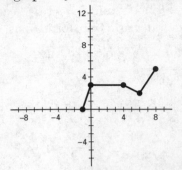

3. $f(x + a)$ will be the graph of $f(x)$ with every point shifted *left* by a units. The graph of $f(x + 2)$ looks like this:

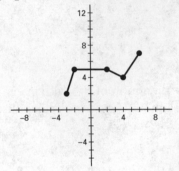

4. $f(x - a)$ will be the graph of $f(x)$ with every point shifted *right* by a units. The graph of $f(x - 2)$ looks like this:

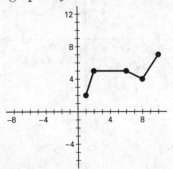

Practice Exercises

1. If a function f is defined as $f = \{(1, 2), (2, 3), (3, 1), (4, 4)\}$, what is $f(2)$?
   (1) 1
   (2) 2
   (3) 3
   (4) 4

2. Which of the following *cannot* be the definition of a function?
   (1) $f = \{(1, 5), (2, 7), (2, 8), (4, 9)\}$
   (2) $f = \{(1, 2), (2, 2), (3, 2), (4, 2)\}$
   (3) $f = \{(0, 0), (1, 1), (-1, 1), (2, 4), (-2, 4)\}$
   (4) $f = \{(6, 1)\}$

3. What is the domain of the function defined as $f = \{(1, 4), (3, 7), (4, 8), (5, 8)\}$?
   (1) $\{4, 7, 8\}$
   (2) $\{1, 3, 4, 5, 7, 8\}$
   (3) $\{1, 3, 4, 5\}$
   (4) $\{4\}$

4. Below is the graph of $y = f(x)$. What is the value of $f(3)$?

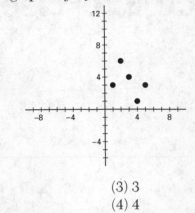

(1) 1   (3) 3

(2) 2   (4) 4

5. Which is the graph of a function?

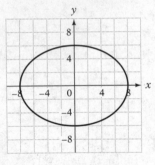

(1)

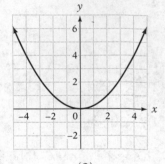

(3)

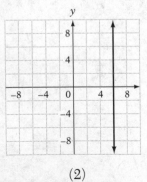

(2)

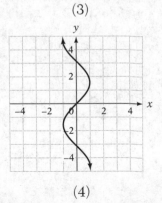

(4)

6. If $g(x) = -x^2 + 7x + 1$ what is $g(2)$?

(1) 11   (3) 27

(2) 19   (4) 35

7. If below is the graph of $y = f(x)$, which is the graph of $y = f(x) - 5$?

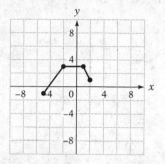

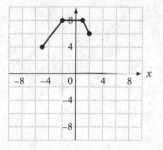

(1)

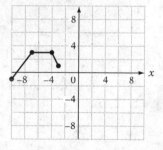

(3)

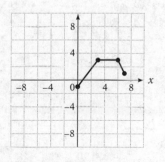

(2)

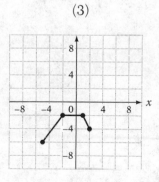

(4)

8. If $f(x) = 5x - 2$, what is $f(x + 1)$?
   (1) $5x - 2$ (3) $5x + 1$
   (2) $5x - 3$ (4) $5x + 3$

9. Below is the graph of $f(x)$ on the left and $g(x)$ on the right. Which is equivalent to $g(x)$?

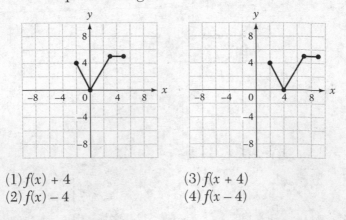

   (1) $f(x) + 4$ (3) $f(x + 4)$
   (2) $f(x) - 4$ (4) $f(x - 4)$

10. If the graph of $y = f(x)$ is a parabola with the vertex at $(5, 1)$, what is the vertex of the graph of the parabola $y = f(x - 2)$?
    (1) $(5, 3)$
    (2) $(5, -1)$
    (3) $(7, 1)$
    (4) $(3, 1)$

Solutions

1. Since the ordered pair $(2, 3)$ is in the set, $f(2) = 3$. The correct choice is **(3)**.

2. A set of ordered pairs cannot be the definition of a function if there are two ordered pairs with the same x-coordinate, but different y-coordinates. In choice 1, there are two ordered pairs with x-coordinates of 2, $(2, 7)$ and $(2, 8)$, so that cannot be the definition of a function. The correct choice is **(1)**.

3. The domain is the set of x-coordinates. In this example, the x-coordinates are 1, 3, 4, and 5. The correct choice is **(3)**.

4. There is a point at $(3, 4)$ so $f(3) = 4$. The correct choice is **(4)**.

5. Choices 1, 2, and 4 all fail the vertical line test since at least one vertical line can pass through more than one point on them. Choice 3 can be the graph of a function since there is no possible vertical line that would pass through more than one point on it. The correct choice is **(3)**.

6. Substitute 2 for x to get $g(2) = -2^2 + 7 \cdot 2 + 1 = -4 + 14 + 1 = 11$. The correct choice is **(1)**. Note that $-2^2 = -4$ and not 4. If there were parentheses around -2, as in $(-2)^2$, then it would equal 4.

7. The graph of $y = f(x) = -5$ is the same as the graph of $y = f(x)$ with each point shifted 5 units down. The correct choice is **(4)**.

8. Substitute $x + 1$ for x to get $f(x + 1) = 5(x + 1) - 2 = 5x + 5 - 2 = 5x + 3$. The correct choice is **(4)**.

9. The graph of $g(x)$ is the same as the graph of $y = f(x)$ with each point shifted 4 units to the right. $g(x)$ must be equivalent to $f(x - 4)$. The correct choice is **(4)**.

10. The graph of $f(x - 2)$ is the same as the graph of $f(x)$ with each point shifted two units to the right. If the vertex of the parabola for $f(x)$ is at $(5, 1)$, the vertex of the parabola for $f(x - 2)$ will be two units to the right of $(5, 1)$, which is at $(7, 1)$. The correct choice is **(3)**.

11. SEQUENCES

11.1 TYPES OF SEQUENCES

A **sequence** is a series of numbers that can be predicted by some kind of pattern. Two of the most common types of sequences are **arithmetic sequences** and **geometric sequences**.

- The sequence 2, 5, 8, 11, 14, … is an example of an arithmetic sequence since each term after the first can be obtained by adding the same number, 3, to the previous term.
- The sequence 2, 6, 18, 54, 162, … is an example of a geometric sequence since each term after the first can be obtained by multiplying the same number, 2, by the previous term.
- There are two notations that are used to describe the terms of a sequence. If the sequence is 2, 5, 8, 11, 14, … the first term can be described as either $a_1 = 2$ or $a(1) = 2$. The second notation is similar to function notation.

11.2 DESCRIBING A SEQUENCE WITH A DIRECT FORMULA

The terms of the sequence 2, 5, 8, 11, … can be described by the formula $a_n = 2 + 3(n - 1)$ or $a(n) = 2 + 3(n - 1)$. If you substitute $n = 1$ into the formula, it becomes $a_1 = 2 + 3(1 - 1) = 2 + 3(0) = 2 + 0 = 2$.

- For any arithmetic sequence, the direct formula for the nth term is $a_n = a_1 + d(n - 1)$ where a_1 is the first term of the sequence and d is the common difference between two consecutive terms. For the sequence 2, 5, 8, 11, …, $a_1 = 2$ and $d = 3$.
- For any geometric sequence, the direct formula for the nth term is $a_n = a_1 \cdot r^{n-1}$, where a_1 is the first term of the sequence and r is the common ratio between two consecutive terms. For the sequence 2, 6, 18, 54, …, $a_1 = 2$ and $r = 3$ since if you divide any term by the previous term you get 3.

11.3 DESCRIBING A SEQUENCE WITH A RECURSIVE FORMULA

Another way to describe the nth term of a sequence is to relate it to the previous term. The term before the a_n term is called the a_{n-1} term. A *recursive formula* for a sequence first defines the exact value of one or more of the terms and then describes how to obtain new terms from the previous ones.

- For the sequence 2, 5, 8, 11, 14, ... the recursive formula is
  $a_1 = 2$
  $a_n = 3 + a_{n-1}$ for $n > 1$

From the set of two equations, the entire sequence can be calculated. For example, when $n = 2$, $a_2 = a_1 + 3 = 2 + 3 = 5$. From this, the value of a_3 can be calculated, and this pattern can continue to obtain other n values. The recursive formulas are not convenient for large values of n, however.

- For the sequence 2, 6, 18, 54, 162, ..., the recursive formula is
  $a_1 = 2$
  $a_n = 3 \cdot a_{n-1}$ for $n > 1$

Practice Exercises

1. What type of sequence is 3, 7, 11, 15, …?
   (1) Increasing arithmetic
   (2) Decreasing arithmetic
   (3) Increasing geometric
   (4) Decreasing geometric

2. What type of sequence is 4, 8, 16, 32, …?
   (1) Increasing arithmetic
   (2) Decreasing arithmetic
   (3) Increasing geometric
   (4) Decreasing geometric

3. What is the next number in the sequence $20, 10, 5, \frac{5}{2}$?

   (1) $\frac{5}{4}$ (3) $\frac{5}{16}$

   (2) $\frac{5}{8}$ (4) $\frac{5}{32}$

4. Find the value of a_2 in the sequence defined by

$$a_1 = 4$$
$$a_n = 3 + a_{n-1} \text{ for } n > 1$$

(1) 3
(2) 5
(3) 7
(4) 12

5. The sequence 5, 11, 17, 23, … can be generated by which definition?

(1) $a_1 = 5$
$a_n = 6a_{n-1} \text{ for } n > 1$

(3) $a_1 = 5$
$a_n = -6 + a_{n-1} \text{ for } n > 1$

(2) $a_1 = 5$
$a_n = \dfrac{11}{5} a_{n-1} \text{ for } n > 1$

(4) $a_1 = 5$
$a_n = 6 + a_{n-1} \text{ for } n > 1$

6. The sequence 3, 15, 75, 375, … can be generated by which definition?

(1) $a_1 = 3$
$a_n = 5\, a_{n-1} \text{ for } n > 1$

(3) $a_1 = 3$
$a_n = \dfrac{1}{5} a_{n-1} \text{ for } n > 1$

(2) $a_1 = 3$
$a_n = 12 + a_{n-1} \text{ for } n > 1$

(4) $a_1 = 3$
$a_n = -12 + a_{n-1} \text{ for } n > 1$

7. A ball is dropped from a window 50 feet above the ground. Each bounce is $\frac{4}{5}$ the height of the previous bounce. Which definition would generate the height of the bounces?

(1) $a_1 = 50$

$a_n = -10 + a_{n-1}$ for $n > 1$

(3) $a_1 = 50$

$a_n = \frac{4}{5} a_{n-1}$ for $n > 1$

(2) $a_1 = 50$

$a_n = 10 + a_{n-1}$ for $n > 1$

(4) $a_1 = 50$

$a_n = \frac{5}{4} a_{n-1}$ for $n > 1$

8. What is the fifth term of the sequence generated by the definition $a_n = 10 - 4(n - 1)$?
   (1) 18
   (2) 12
   (3) 2
   (4) –6

9. What definition would produce the sequence 4, 13, 22, 31, …?
   (1) $a_n - 4 + 9n$
   (2) $a_n = 4 + 9(n - 1)$
   (3) $a_n = 9 + 4n$
   (4) $a_n = 9 + 4(n - 1)$

10. Which expression could be used to find the 20th term of the sequence 5, 9, 13, 17, 21, …?
    (1) $5 + 4(20)$
    (2) $5 + 4(19)$
    (3) $4 + 5(20)$
    (4) $4 + 5(19)$

Solutions

1. Since $7 = 3 + 4$ and $11 = 7 + 4$ and $15 = 11 + 4$, this is an increasing arithmetic sequence. The correct choice is **(1)**.

2. Since $8 = 2 \cdot 4$ and $16 = 2 \cdot 8$ and $32 = 2 \cdot 16$, this is an increasing geometric sequence. The correct choice is **(3)**.

3. Each term is equal to $\frac{1}{2}$ multiplied by the previous term. So the next term is $\frac{1}{2} \cdot \frac{5}{2} = \frac{5}{4}$. The correct choice is **(1)**.

4. $a_2 = 3 + a_{2-1} = 3 + a_1 = 3 + 4 = 7$. The correct choice is **(3)**.

5. The first term is a_1, so $a_1 = 5$. Each term is equal to 6 plus the previous term, so $a_n = 6 + a_{n-1}$. The correct choice is **(4)**.

6. The first term is a_1, so $a_1 = 3$. Each term is equal to 5 multiplied by the previous term, so $a_n = 5 \cdot a_{n-1}$. The correct choice is **(1)**.

7. The initial height is 50, so $a_1 = 50$. Each bounce is 4/5 the height of the previous bounce so $a_n = 4/5 \cdot a_{n-1}$. The correct choice is **(3)**.

8. Substitute 5 for n to get $a_5 = 10 - 4(5 - 1) = 10 - 4(4) = 10 - 16 = -6$. The correct choice is **(4)**.

9. Substitute 1 for n in each equation. Only the second equation gets $a_1 = 4$. Also, an arithmetic sequence has the form $a_n = a_1 + d(n - 1)$, where d is the common difference and a_1 is the first term. Since $a_1 = 4$ and $d = 9$, the equation is $a_n = 4 + 9(n - 1)$. The correct choice is **(2)**.

10. An arithmetic sequence with first term a_1 and common difference of d has the equation $a_n = a_1 + (n - 1)d$. a_1 is 5 in this example, and $d = 4$. Therefore, the equation is $a_n = 5 + 4(n - 1)$. Substitute 20 for n to get a $20 = 5 + 4(19)$. The correct choice is **(2)**.

12. REGRESSION CURVES

12.1 THE LINE OF BEST FIT

A **line of best fit** is a line that comes as close as possible to a set of points on a graph. In the scatterplot below, there is no line that could pass through all ten points. Of all the possible lines, though, there is one that is a better fit than the others and this is the line of best fit.

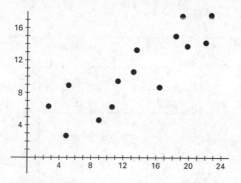

The equation for the line of best fit can be determined quickly on a graphing calculator.

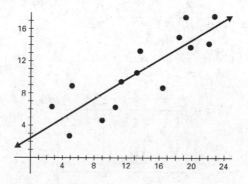

Below is a scatter plot and the chart on which it was based.

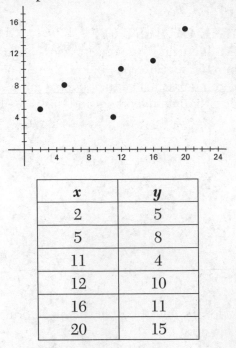

x	y
2	5
5	8
11	4
12	10
16	11
20	15

Instructions for the TI-84:

Press [STAT] and [1], and enter the x values into L1 and the y values into L2.

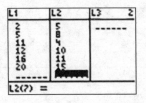

Press [STAT], [right], [4], and [ENTER] for the equation of the line of best fit.

Instructions for the TI-Nspire:

From the home screen select the Add Lists and Spreadsheet icon. In the header row, label column A with an x and column B with a y. Enter the x values into cells A1 to A6. Enter the y values into cells B1 to B6.

Press [menu], [4], [1], and [3] and select x in the X List field and y in the Y List field.

Press the OK button.

< 1.1 >		*Unsaved ▼		
A x	B y	C	D	
◆			=LinRegM	
1	2	5	Title	Linear Re..
2	5	8	RegEqn	m*x+b
3	11	4	m	0.477679
4	12	10	b	3.57887
5	16	11	r²	0.617042
D1	="Linear Regression (mx+b)"			< >

The equation for the line of best fit is $y = 0.477679x + 3.57887$.

12.2 THE CORRELATION COEFFICIENT

The **correlation coefficient**, denoted by the variable r, is a number between -1 and 1. When a line of best fit has a positive slope and passes exactly through each of the points, the correlation coefficient is 1. When a line of best fit has a negative slope and passes exactly through each of the points, the correlation coefficient is -1. All correlation coefficients are between -1 and 1.

- The line of best fit for the scatter plot below is close to 1 because it has a positive slope and comes close to the points on the scatter plot. In this case $r = 0.9$

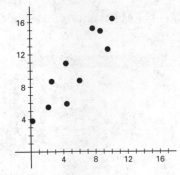

- The line of best fit for the following scatter plot is close to -1 because it has a negative slope and comes close to the points on the scatter plot, but not as close as the line in the previous example did to the points on that graph. In this case the r = –0.8

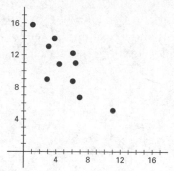

The graphing calculator can display the correlation coefficient. On the TI-84 it will be displayed along with the line of best fit only if the "diagnostics" are turned on. Press [2ND] and [0] to access the catalog and scroll to the DiagnosticOn command to do this. On the TI-Nspire it will display the correlation coefficient along with the equation of the line of best fit.

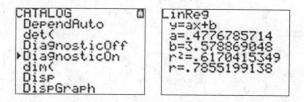

The correlation coefficient for the example from Section 12.1 was $r = 0.78552$.

12.3 RESIDUAL PLOTS

When a line is a good fit for a scatterplot, the points in the plot are close to the points on the line. One way to measure this is to calculate the *residuals* and to examine them on a *residual plot*.

Here is a scatterplot with five points $(1, 2)$, $(2, 4)$, $(3, 4)$, $(4, 6)$, $(5, 7)$.

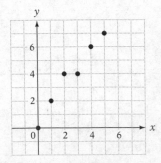

The line of best fit for this scatterplot is $y = 1.2x + 1$.

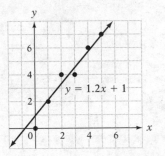

Only one of the points is on the line. For the others, two are above the line and two are below. Residuals measure how far above or below the line the points are. The information can be collected on a chart.

x	y	$1.2x + 1$	Residuals (y column) − ($1.2x + 1$ column)
1	2	2.2	−0.2
2	4	3.4	0.6
3	4	4.6	−0.6
4	6	5.8	0.2
5	7	7	0

The two points above the line have positive residuals. The two points below the line have negative residuals.

A residual plot is a scatter plot with the x-coordinates from the x column in the chart and the y-coordinates from the residuals column in the chart.

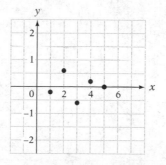

MATH FACTS

When the original scatterplot resembles a line, the residual plot will be a random scattering of points with no apparent pattern. When the original scatterplot resembles some other kind of curve, the residual plot will not be a bunch of random points but will look like a line or some kind of curve like a line or a U shape.

Example 1
Construct a residual plot for the scatter plot based on the following chart. What does this residual plot suggest about the original scatterplot?

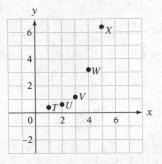

Solution: The line of best fit is $y = 1.453x - 1.987$.

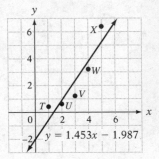

x	y	$1.453x - 1.987$	Residuals (y column) $- (1.453x - 1.987)$
1	0.42	−0.534	0.954
2	0.62	0.919	−0.299
3	1.22	2.372	−1.152
4	3.21	3.825	−0.615
5	6.39	5.278	1.112

The residual plot looks like this:

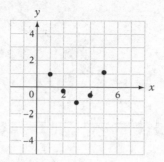

Because this has a U shape and is not just a random scattering of points, it suggests that the original scatter plot did not resemble a line.

12.4 PARABOLAS AND EXPONENTIALS OF BEST FIT

Even though every scatter plot has a line of best fit, sometimes even the best possible line isn't a very good fit.

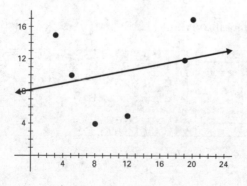

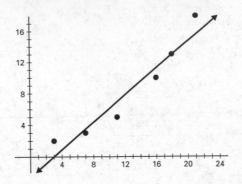

If the points in a scatter plot look more like a parabola or an exponential curve, the most appropriate curve to model it may not be a line. For parabolas, there is a parabola of best fit, and for exponential curves, there is an exponential curve of best fit. Both can be calculated on the graphing calculator. Select quadratic regression for the parabola of best fit and exponential regression for the exponential curve of best fit.

For the TI-84:

Enter the data into L1 and L2. Then press [STAT], [right], and [5].

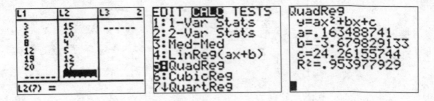

For the TI-Nspire:

To find the parabola of best fit, after entering the data into column A and column B, press [menu], [4], [1], and [6].

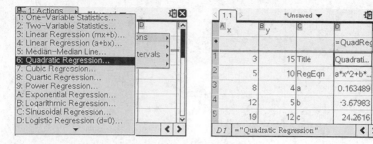

To find the exponential curve of best fit on the TI-84, enter the data into L1 and L2 and press [STAT], [right], and [0].

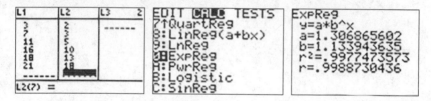

To find the exponential curve of best fit on the TI-Nspire, enter the data into column A and column B, press [menu], [4], [1], and [A].

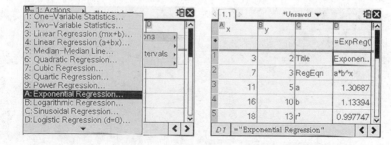

The parabola of best fit for the first scatter plot is $y = 0.1634989x^2 - 3.67983x + 24.2616$.

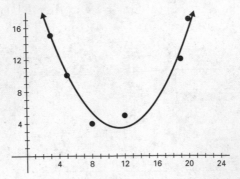

The exponential curve of best fit for the second scatter plot is $y = 1.30687 \cdot 1.13394^x$.

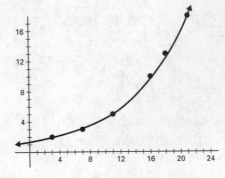

Practice Exercises

1. Calculate the equation for the line of best fit for the following set of data in $y = mx + b$ form. Round m and b to the nearest tenth.

x	y
1	3
2	5
3	4
4	6
5	8

(1) $y = 1.1x + 1.9$
(2) $y = 1.4x + 1.7$
(3) $y = 1.7x + 1.4$
(4) $y = 1.9x + 1.1$

2. Calculate the equation for the line of best fit for the following set of data in $y = mx + b$ form. Round m and b to the nearest tenth.

x	y
10	33
20	20
30	10
40	14
50	6

(1) $y = 34.6x - 0.6$ (3) $y = 31.8x - 0.7$
(2) $y = -0.6x + 34.6$ (4) $y = -0.7x + 31.8$

3. What is the equation for the line of best fit for the points on this scatter plot?

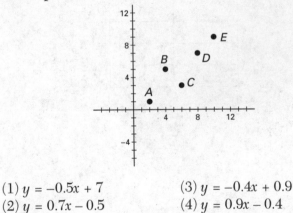

(1) $y = -0.5x + 7$ (3) $y = -0.4x + 0.9$
(2) $y = 0.7x - 0.5$ (4) $y = 0.9x - 0.4$

4. Of these four choices, which line appears to be the best fit for this scatter plot?

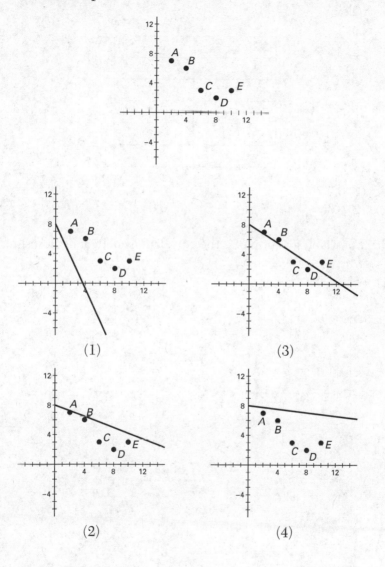

(1)

(2)

(3)

(4)

5. What is the correlation coefficient (r), rounded to the nearest hundredth, for the line of best fit for the data on the table below?

x	y
3	10
6	13
9	27
12	38
15	40

(1) 0.97   (3) 0.95
(2) 0.96   (4) 0.94

6. For which scatter plot is the correlation coefficient closest to 1?

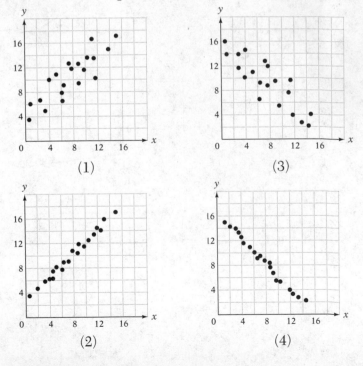

(1)   (3)

(2)   (4)

7. Of the four choices, which is closest to the correlation coefficient for this scatter plot?

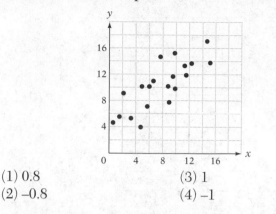

(1) 0.8   (3) 1
(2) –0.8   (4) –1

8. Find the equation of the parabola of best fit for the data on the table below.

x	y
1	2
2	2
3	3
4	4
5	6
6	8
7	10
8	14

(1) $y = 0.31x^2 - 67x + 3.53$
(2) $y = 0.24x^2 - 0.52x + 2.23$
(3) $y = 0.58x^2 - 0.41x + 4.12$
(4) $y = 0.73x^2 - 0.72x + 6.87$

9. Find the equation of the exponential curve of best fit for the data on the table below.

x	y
1	2
2	2
3	3
4	4
5	6
6	8
7	10
8	14

(1) $y = 1.28 \cdot 1.35^x$ (3) $y = 1.47 \cdot 1.39^x$

(2) $y = 1.35 \cdot 1.28^x$ (4) $y = 1.39 \cdot 1.47^x$

10. For which scatter plot would an exponential curve of best fit be most appropriate?

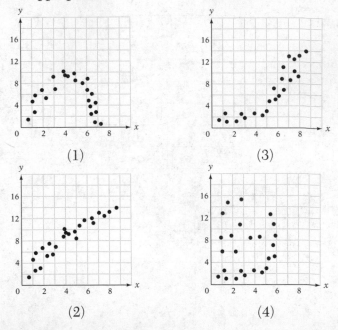

(1)   (3)

(2)   (4)

Solutions

1. Enter the data into the graphing calculator and do linear regression to get the equation $y = 1.1x + 1.9$. The correct choice is (1).

2. Enter the data into the graphing calculator and do linear regression to get the equation $y = -0.6x + 34.6$. The correct choice is (2).

3. Enter 2, 4, 6, 8, 10 for the x values and 1, 5, 3, 7, 9 for the y values into the graphing calculator and do linear regression. The equation is $y = 0.9x - 0.4$. The correct choice is (4).

4. Of the four choices, choice 3 seems to have the line closer to the points than the other three choices. The correct choice is (3).

5. Enter the data into the graphing calculator and do linear regression to get an r value of 0.97. The correct choice is (1).

6. To have an r value close to +1, the points must lie close to a line with a positive slope. Choices 1 and 2 both resemble lines with positive slopes. In choice 2 the points seem to fall closer to a straight line than choice 1. The correct choice is (2).

7. Since this scatter plot resembles a line with a positive slope, choices 2 and 4 can be eliminated because r must be positive. To have an r value of +1 the points would have to lie perfectly on a line, so choice 3 can be eliminated too. The correct choice is (1).

8. Enter the data into the graphing calculator and do quadratic regression to get $y = 0.24x^2 - 0.52x + 2.23$. The correct choice is (2).

9. Enter the data into the graphing calculator and do exponential regression to get $y = 1.28 \cdot 1.35^x$. The correct choice is (1).

10. Of the four choices, choice 3 looks the most like an exponential curve. Choice 1 looks like a parabola. Choice 2 looks like a line. Choice 4 looks like a bunch of random points. The correct choice is (3).

13. STATISTICS

13.1 MEAN, MEDIAN, AND MODE

In a set of numbers, the **mean**, also known as the **average**, is the sum of the numbers divided by how many numbers there are. For the set 70, 75, 78, 80, 82, the mean can be calculated as $\frac{70+75+78+80+82}{5} = \frac{385}{5} = 77$.

The **median** is the middle number (or average of the two middle numbers if there are an even amount of numbers) when the numbers are arranged from least to greatest. In the set 70, 75, 78, 80, 82, the median is the number 78. If the set were 70, 75, 78, 80, 82, 85, the median would be the average of 78 and 80, which is $\frac{78+80}{2} = 79$.

The **mode** is the number that appears most frequently. In the set 70, 75, 78, 80, 80, the mode is 80 since there are two 80s and only one of each of the other numbers.

13.2 FIRST QUARTILE AND THIRD QUARTILE

The median is greater than 50% of the numbers in the list. The number that is greater than just 25% of the numbers on the list is called the *first quartile*. The number that is greater than 75% of the numbers on the list is called the *third quartile*. To find the first quartile, find the median of all the numbers less than the median of the list. To find the third quartile, find the median of all the numbers greater than the median of the list. The interquartile range is the difference between the third quartile and the first quartile.

- For the list 40, 43, 45, 47, 48, 52, 57, 60, 61, 64, 68:
- The numbers less than the median 52 are 40, 43, 45, 47, 48. The median of these five numbers is 45, which is the first quartile.
- The numbers greater than the median 52 are 57, 60, 61, 64, 68. The median of these five numbers is 61, which is the third quartile. The interquartile range is $61 - 45 = 16$.

13.3 BOX PLOTS

A box plot is a picture that shows five different metrics, minimum, first quartile, median, third quartile, and maximum. To make a box plot, draw a segment connecting the maximum and the minimum. Then draw a rectangle with sides at the first quartile and third quartile. Finally, draw a line through the rectangle at the median.

- For the following numbers, the box plot looks like this:

10, 10, 10, 12, 12, 12, 12, 17, 25, 25, 25, 25, 30, 30, 30

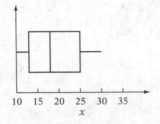

13.4 USING THE GRAPHING CALCULATOR TO DETERMINE MAXIMUM, MINIMUM, MEDIAN, FIRST QUARTILE, AND THIRD QUARTILE

For the TI-84:

The graphing calculator can calculate the five measures of central tendency. First enter all the numbers into L1 by pressing [STAT] and [1] for Edit.

To find the minimum, first quartile, median, third quartile, and maximum of the seven numbers 10, 4, 8, 12, 6, 16, 14, enter them into L1. Then press [STAT] and [1] for 1-Var Stats and press [ENTER].

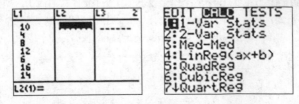

The screen will display

```
1-Var Stats
x̄=10
Σx=70
Σx²=812
Sx=4.320493799
σx=4
↓n=7
```

The $\bar{x} = 10$ is for the mean. The $n = 7$ means that there were seven elements in the list. For the minimum, first quartile, median, third quartile, and maximum, press the down arrow five times.

```
1-Var Stats
↑n=7
minX=4
Q1=6
Med=10
Q3=14
maxX=16
```

minX is for the minimum, Q1 is for the first quartile, Med is for the median, Q3 is for the third quartile, and maxX is for the maximum element.

For the TI-Nspire:

From the home screen, select the Add Lists & Spreadsheet icon. Name column A x and fill in cells A1 through A7 with the numbers 10, 4, 8, 12, 6, 16, 14.

Press [4], [1], and [1] for One-Variable Statistics.

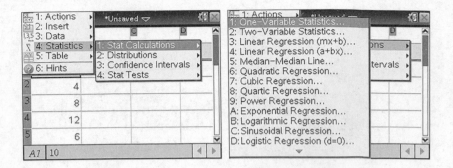

Press [OK] since there is just one list. Set the X1 List to x since that was what the column with the data was named in the spreadsheet.

In cells C2 through C13, the one-variable statistics will be displayed. The median is the $\bar{x}$-bar. n is for the number of numbers. MinX is the smallest number. Q1X is the first quartile. MedianX is the median. Q3X is the third quartile. MaxX is the largest number.

Practice Exercises

1. Find the mean of this set of numbers {4, 5, 8, 8, 8, 10, 10, 13, 15, 17, 23}.
   (1) 8   (3) 10
   (2) 9   (4) 11

2. Find the median of this set of numbers {4, 5, 8, 8, 8, 10, 10, 13, 15, 17, 23}.
   (1) 8   (3) 11
   (2) 10   (4) 15

3. Find the interquartile range of this set of numbers {4, 5, 8, 8, 8, 10, 10, 13, 15, 17, 23}.
   (1) 7   (3) 9
   (2) 8   (4) 10

4. For the first four days of a five-day vacation, the mean temperature was 80 degrees. What must the temperature be on the fifth day in order for the mean temperature to be 82 degrees?
   (1) 88   (3) 90
   (2) 89   (4) 91

5. For which data set is the median greater than the mean?
   (1) {4, 7, 10, 13, 16}   (3) {8, 9, 10, 11, 12}
   (2) {8, 9, 10, 18, 19}   (4) {1, 2, 10, 11, 12}

6. What is the mode of the data in this histogram?

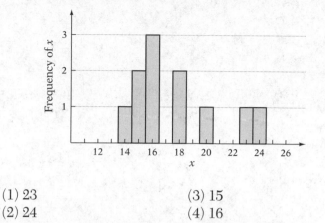

(1) 23   (3) 15
(2) 24   (4) 16

7. What is the median of the data in this box plot?

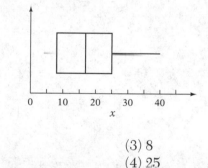

(1) 17   (3) 8
(2) 15.2   (4) 25

8. What is true about this data set: 1, 2, 10, 11, 11?
(1) The median is greater than the mean.
(2) The median is equal to the mean.
(3) The median is less than the mean.
(4) The median is greater than the mode.

9. For which data set is the interquartile range equal to 0?
   (1) 2, 3, 6, 6, 6, 7, 8
   (2) 2, 6, 6, 6, 6, 6, 8
   (3) 1, 2, 3, 6, 7, 8, 9
   (4) 2, 6, 6, 6, 6, 7, 8

10. In a set of seven numbers, the largest number is increased by 10. Which measure of central tendency must increase because of this?
    (1) Mean
    (2) Mode
    (3) First quartile
    (4) Interquartile range

Solutions

1. To get the mean, add the 11 values and divide by 11. $(4 + 5 + 8 + 8 + 8 + 10 + 10 + 13 + 15 + 17 + 23)/11 = 121/11 = 11$. The correct choice is (**4**).

2. The median is the middle number after the numbers have been arranged from least to greatest. These numbers are already arranged from least to greatest so the middle number is the sixth number, which is 10. The correct choice is (**2**).

3. The interquartile range is the difference between the first quartile and the third quartile. The median is the sixth number so the first quartile is the median of the first five numbers. The first five numbers are 4, 5, 8, 8, 8 with a median of 8. The third quartile is the median of the last five numbers. The last five numbers are 10, 13, 15, 17, 23, with a median of 15. The first quartile is 8. The third quartile is 15. The interquartile range is $15 - 8 = 7$. The correct choice is (**1**).

4. The sum of a set of numbers is the product of the average and the number of numbers. If the first four days had an average of 80, the sum of all four temperatures was 320. If the first five days need to have an average of 82, the sum of all five temperatures needs to be $82 \cdot 5 = 410$. The fifth temperature needs to be $410 - 320 = 90$. The correct choice is (**3**).

5. The median for all four data sets is 10. The mean of the numbers in choice 4 is $(1 + 2 + 10 + 11 + 12)/5 = 7.2$, which is less than the median. All the other choices have the mean greater than or equal to the median. The correct choice is (**4**).

6. The mode is the most frequent number. In a histogram the most frequent number is the tallest bar, which is 16 in this example. The correct choice is (**4**).

7. The vertical line in the middle of the rectangle represents the median in a box plot. Since this is at 17, the median is 17. The correct choice is (**1**).

8. The median of the data set is 10. The mode is 11. The mean is 7. The median is greater than the mean. The correct choice is **(1)**.

9. The interquartile range is zero if the first quartile is equal to the third quartile. This is the case for choice 2 where both the first quartile and the third quartile are 6. The correct choice is **(2)**.

10. When the largest number is increased by 10, the sum of all the numbers is increased by 10 so the mean will increase. The other measures will not be affected by increasing the largest number. The correct choice is **(1)**.

Glossary of Terms

Addition property of equality A property of algebra that states that when equal values are added to both sides of a true equation, the equation continues to be true. To solve the equation $x - 2 = 5$, add 2 to both sides of the equation by using the addition property of equality.

Arithmetic sequence A number sequence in which the difference between two consecutive terms is a constant. The sequence 2, 5, 8, 11, 14, … is an arithmetic sequence because the difference between consecutive terms is always 3.

Axis of symmetry An imaginary vertical line that passes through the vertex of a parabola. The equation for the axis of symmetry of a parabola is defined by $y = ax^2 + bx + c$ is $x = \dfrac{-b}{2a}$.

Base The number being raised to a power in an exponential expression. In the expression $2 \cdot 3^x$, the 3 is the base.

Binomial A polynomial with only two terms. $3x + 5$ is a binomial.

Box plot A graphical way to summarize data. The five numbers represented by the minimum, first quartile, median, third quartile, and maximum are graphed on a number line. A line segment connects the minimum to the first quartile. A rectangle is drawn around the first quartile and third quartile with a vertical line at the median. A line segment connects the third quartile to the maximum.

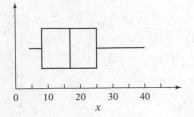

Closed form defined sequence A formula that defines the nth term of a sequence. The formula $a_n = 3 + 2(n - 1)$ is a closed form definition of a sequence. To get the 50th term of the sequence, substitute 50 for n in the definition.

Coefficient A number multiplied by a variable expression. In the expression $5x + 2$, 5 is the coefficient of x.

Common difference In an arithmetic sequence, the difference between consecutive terms. The common difference in the sequence 2, 5, 8, 11, 14, … is 3.

Common ratio In a geometric sequence, the ratio between consecutive terms. The common ratio in the sequence 2, 6, 18, 54, 162, … is 3 since $162/54 = 54/18 = 18/6 = 6/2 = 3$.

Commutative property of addition or multiplication The law from arithmetic that states that the order in which two numbers are added or multiplied does not matter. Because of the commutative property of addition, $5 + 2 = 2 + 5$.

Completing the square A method of solving a quadratic equation that involves turning one side of the equation into a perfect square trinomial.

Constant A number that does not have a variable part. In the expression $5x + 2$, 2 is a constant.

Correlation coefficient A number represented by r that measures how well a curve of best fit matches the points in a scatter plot.

When the correlation coefficient is very close to 1 or to –1, the curve is a very good fit.

Degree of a polynomial The exponent on the highest power of a polynomial. In the polynomial $x^3 - 2x^2 + 5x - 2$, the degree is 3 since the highest power is a 3.

Difference of perfect squares Factoring a quadratic binomial in the form $x^2 - a^2$ into $(x - a)(x + a)$. For example, $x^2 - 9 = (x - 3)(x + 3)$.

Distributive property of multiplication over addition The rule that allows expressions of the form $a(b + c)$ to become $a \cdot b + a \cdot c$. For example, $2(3x + 5) = 6x + 10$.

Division property of equality A property of algebra that states that when both sides of a true equation are divided by the same non-zero number, the equation continues to be true. To solve the equation $2x = 8$, divide both sides of the equation by 2 using the division property of equality.

Domain The numbers that can be input into a function. When the function is defined as a set of ordered pairs or as a graph, the domain is the set of x-coordinates.

Dot plot A graphical way of representing a data set where each piece of data is represented with a dot.

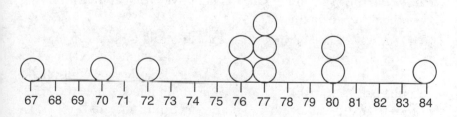

Elimination method A way of solving a system of linear equations by combining the two equations in such a way as to eliminate one of the variables.

In the set of equations,

$$x + 2y = 12$$
$$3x - 2y = -4$$

the y variable is eliminated by adding the two equations together.

Equation Two mathematical expressions with an equals sign between them. $3x + 2 = 8$ is an equation.

Exponential equation An equation in which the variable is an exponent. $2 \cdot 3^x = 18$ is an exponential equation.

Exponential function A function in which the variable is an exponent. $f(x) = 2 \cdot 3^x$ is an exponential function.

Expression Numbers and variables that are combined with the operations from math.

Factoring a polynomial Finding two polynomials that can be multiplied to become another polynomial. The polynomial $x^2 + 5x + 6$ can be factored into $(x + 2)(x + 3)$.

Factors The polynomials that evenly divide into a polynomial. The factors of $x^2 + 5x + 6$ are $(x + 2)$ and $(x + 3)$.

First quartile The number in a data set that is bigger than just 25% of the numbers in the set.

FOIL A way of multiplying two binomials of the form $(a + b)(c + d)$ where (F)irst the a and c are multiplied, then the (O)uters a and d are multiplied, then the (I)nners b and c are multiplied, and finally the (L)asts b and d are multiplied. Then the four results are added together. The product of $(x + 2)$ and $(x + 3)$ is $x^2 + 3x + 2x + 6 = x^2 + 5x + 6$ by this method.

Function Something that takes numbers as inputs and outputs numbers. Functions are often labeled with the letters f or g. The notation $f(2) = 7$ means that when the number 2 is input into function f, it outputs the number 7.

Geometric sequence A number sequence in which the ratio between two consecutive terms (what you get when you divide one term by the term before it) is a constant. The sequence 2, 6, 18, 54, 162, ... is a geometric sequence because the ratio between consecutive terms is always 3.

Graph A visual way to describe the solution set to an equation. Each solution to the equation corresponds to an ordered pair that is graphed as a point on the coordinate plane. Each of the ordered pairs that satisfy an equation produce a point on the coordinate plane and the collection of all the points is the graph of the equation.

Greatest common factor The largest expression that divides evenly into two or more monomials. The greatest common factor of $6x^2$ and $8x^3$ is $2x^2$.

Growth rate In an exponential expression $a \cdot b^x$, the b is the growth rate. For example, in the equation $y = 500 \cdot 1.05^x$, the growth rate is 1.05.

Histogram A way of representing data with repeated values. Each value is represented by a bar whose height corresponds to the number of times that value is repeated. There are no spaces between the bars.

Increasing A function is increasing on an interval if making the input value larger also makes the y output larger. On a graph, an increasing function "goes up" from left to right.

Inequality Like an equation, but there is a $<$, $>$, $\leq$, or $\geq$ sign between the two expressions. $x + 2 > 5$ in a one-variable inequality. $y \leq 2x + 6$ is a two-variable inequality.

Interquartile range The difference between the number that is the third quartile of a data set and the number that is the first quartile of a data set.

Isolating a variable A variable is isolated when it is by itself on one side of an equation. In the equation $x + 2 = 5$, the x is not yet isolated. Subtracting 2 from both sides of the equation transforms the original equation into $x = 3$ with the x now isolated.

Like terms Terms that have the same variable part. They can be combined by adding or subtracting. $2x^2$ and $3x^2$ are like terms. $2x^2$ and $3x^3$ are not like terms.

Line of best fit A line that comes closest to the set of points in a scatter plot.

Linear equation An equation in which the greatest exponent is a 1. The equation $2x + 3 = 7$ is a linear equation.

Linear function A function in which the greatest exponent is a 1. The function $f(x) = 2x + 3$ is a linear function.

Mean The average of the numbers in a data set. Calculate the mean by adding all the numbers and dividing the total by the number of numbers in the set.

Median The middle number in a data set after it has been arranged from least to greatest. If there are an even number of numbers in the data set, the median is found by adding the two middle numbers and dividing by 2.

Mode The most frequent number in a data set.

Monomial A mathematical expression that has a coefficient and/or a variable part. $3x^2$ is a monomial. $3 + x^2$ is not a monomial.

Multiplication property of equality A property of algebra that states that when equal values are multiplied by both sides of a true equation, the equation continues to be true. To solve the equation $(1/2)x = 5$, multiply both sides of the equation by 2, using the multiplication property of equality.

Ordered pair Two numbers written in the form (x, y). An ordered pair can be a solution to a two-variable equation. For example, $(2, 5)$ is one solution to the equation $y = 2x + 1$. Ordered pairs can be graphed on the coordinate axes by locating the point with the x-coordinate equal to the x value and the y-coordinate equal to the y value.

Parabola A "U"-shaped curve that is the graph of the solution set of a quadratic equation.

Perfect square trinomial A quadratic trinomial of the form $x^2 + bx + (b/2)^2$, which can be factored into $(x + b/2)^2$. For example, $x^2 + 6x + 9 = (x + 3)^2$.

Piecewise function A function that has multiple rules for determining output values from input values, depending on what the input values are. If the function $f(x)$ is defined as

$$f(x) = 2x + 1 \qquad \text{if } x < 0$$
$$x^2 \qquad \text{if } x \geq 0$$

then $f(-3) = 2(-3) + 1 = -1$ and $f(5) = 5^2 = 25$.

Polynomial The sum of one or more monomials. Each term of the polynomial has the form ax^n. $3x^4 + 2x^3 - 3x^2 + 6x - 1$ is a polynomial.

Quadratic equation An equation in which the highest power on a variable is a 2. $x^2 + 5x + 6 = 0$ is a quadratic equation.

Quadratic formula The formula $x = \left(-b \pm \sqrt{(b^2 - 4ac)}\right)/2a$. This formula can be used to find the two solutions to the quadratic equation $ax^2 + bx + c = 0$.

Quadratic function A function in which the highest power on a variable is 2. $f(x) = x^2 + 5x + 6$ is a quadratic function.

Quadratic polynomial A polynomial in which the highest power on a variable is 2. $x^2 + 5x + 6$ is a quadratic polynomial.

Range The set of values that can be output from a function is the range of that function.

Recursively defined sequence A way to define a sequence in which the first term or terms of the sequence are given and a formula is given for calculating the next term based on the previous term or terms.

$$a_1 = 5$$
$$a_n = 3 + a_{n-1} \text{ for } n > 1$$

is a recursive definition for the sequence 5, 8, 11, 14, ...

Regression Finding a curve that best fits a scatterplot. Three types of regression are linear, quadratic, and exponential.

Residual plot A set of points that represent how far points on a graph deviate from a curve of best fit.

Roots The roots of an equation are the values that solve that equation. The roots of $x^2 + 5x + 6 = 0$ are −3 and −2.

Sequence A list of numbers that usually have some kind of pattern.

Slope The slope of a line is a measure of the line's steepness. The equation for the slope of a line that passes through the two points (x_1, y_1) and (x_2, y_2) is $m = \dfrac{y_2 - y_1}{x_2 - x_1}$.

Slope-intercept form An equation in the form $y = mx + b$ where m and b are numbers in slope intercept form. $y = 2x - 1$ is in slope intercept form. When a two-variable equation is in slope intercept form, the graph of the equation has a y-intercept of $(0, b)$ and a slope of m.

Solution set The set of numbers or ordered pairs that satisfies an equation. The solution set of $x + 2 = 5$ is {3}. The solution set of $x + y = 10$ has an infinite number of ordered pairs in its solution set, including (2, 8), (3, 7), and (4, 6).

Substitution method A method for solving a system of equations in which one variable is isolated in one of the equations, and the expression equal to that variable is substituted for it in the other equation.

Subtraction property of equality A property of algebra that states that when equal values are subtracted from both sides of a true equation, the equation continues to be true. To solve the equation $x + 2 = 5$, subtract 2 from both sides of the equation by using the subtraction property of equality.

System of equations Two or more equations with two or more unknowns to solve for. An example of a system of equations with a solution of $(8, 2)$ is

$$x + y = 10$$
$$x - y = 6$$

Third quartile The number in a data set that is greater than just 75% of the numbers in the set.

Translation A translation of a graph is when the points on it are each shifted the same amount in the same direction. Examples of translations are vertical translations, horizontal translations, and combinations of vertical and horizontal translations.

Trinomial A polynomial with three terms. The polynomial $x^2 + 5x + 6$ is a trinomial.

Variable A letter, often an x, y, or z, that represents a value in a mathematical expression. In an algebraic equation, the variable is often the unknown that needs to be solved for.

Vertex The turning point of a parabola is its vertex. If the parabola opens upward, the vertex is the minimum point. If the parabola opens downward, the vertex is the maximum point.

Vertical line test A way of testing to see if a graph can represent a function. If at least one vertical line can pass through at least two points on the graph, the graph fails the vertical line test and cannot represent a function. If there are no vertical lines that can pass through at least two points, then the graph can be the graph of a function.

x-intercept The location where a curve crosses the x-axis. The y-coordinate of the x-intercept is 0.

y-intercept The location where a curve crosses the y-axis. The x-coordinate of the y-intercept is 0. In slope-intercept form, $y = mx + b$, the y-intercept is located at $(0, b)$.

Zeros The zeros of a function f are the numbers that can be input into the function so that 0 is output from the function. For example, the function $f(x) = 2x - 6$ has the number 3 as its only zero since $f(3) = 2(3) - 6 = 6 - 6 = 0$.

Regents Examinations, Answers, and Self-Analysis Charts

Examination
August 2015
Algebra I

HIGH SCHOOL MATH REFERENCE SHEET

Conversions

1 inch = 2.54 centimeters	1 cup = 8 fluid ounces
1 meter = 39.37 inches	1 pint = 2 cups
1 mile = 5280 feet	1 quart = 2 pints
1 mile = 1760 yards	1 gallon = 4 quarts
1 mile = 1.609 kilometers	1 gallon = 3.785 liters
	1 liter = 0.264 gallon
1 kilometer = 0.62 mile	1 liter = 1000 cubic centimeters
1 pound = 16 ounces	
1 pound = 0.454 kilogram	
1 kilogram = 2.2 pounds	
1 ton = 2000 pounds	

Formulas

Triangle	$A = \dfrac{1}{2}bh$
Parallelogram	$A = bh$
Circle	$A = \pi r^2$
Circle	$C = \pi d$ or $C = 2\pi r$

Formulas (continued)

General Prisms	$V = Bh$
Cylinder	$V = \pi r^2 h$
Sphere	$V = \dfrac{4}{3}\pi r^3$
Cone	$V = \dfrac{1}{3}\pi r^2 h$
Pyramid	$V = \dfrac{1}{3}Bh$
Pythagorean Theorem	$a^2 + b^2 = c^2$
Quadratic Formula	$x = \dfrac{-b \pm \sqrt{b^2 - 4ac}}{2a}$
Arithmetic Sequence	$a_n = a_1 + (n-1)d$
Geometric Sequence	$a_n = a_1 r^{n-1}$
Geometric Series	$S_n = \dfrac{a_1 - a_1 r^n}{1 - r}$ where $r \neq 1$
Radians	$1 \text{ radian} = \dfrac{180}{\pi} \text{ degrees}$
Degrees	$1 \text{ degree} = \dfrac{\pi}{180} \text{ radians}$
Exponential Growth/Decay	$A = A_0 e^{k(t - t_0)} + B_0$

PART I

Answer all 24 questions in this part. Each correct answer will receive 2 credits. No partial credit will be allowed. For each statement or question, write in the space provided the numeral preceding the word or expression that best completes the statement or answers the question.   [48 credits]

1 Given the graph of the line represented by the equation $f(x) = -2x + b$, if b is increased by 4 units, the graph of the new line would be shifted 4 units

(1) right   (3) left

(2) up   (4) down   1 __2__

2 Rowan has $50 in a savings jar and is putting in $5 every week. Jonah has $10 in his own jar and is putting in $15 every week. Each of them plots his progress on a graph with time on the horizontal axis and amount in the jar on the vertical axis. Which statement about their graphs is true?

(1) Rowan's graph has a steeper slope than Jonah's.

(2) Rowan's graph always lies above Jonah's.

(3) Jonah's graph has a steeper slope than Rowan's.

(4) Jonah's graph always lies above Rowan's.   2 __3__

3 To watch a varsity basketball game, spectators must buy a ticket at the door. The cost of an adult ticket is $3.00 and the cost of a student ticket is $1.50. If the number of adult tickets sold is represented by a and student tickets sold by s, which expression represents the amount of money collected at the door from the ticket sales?

(1) $4.50as$ (3) $(3.00a)(1.50s)$

(2) $4.50(a + s)$ (4) $3.00a + 1.50s$ 3 __4__

4 The graph of $f(x)$ is shown below.

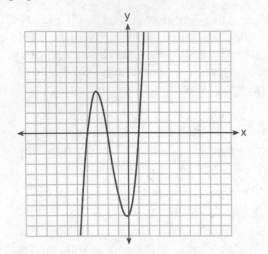

Which function could represent the graph of $f(x)$?

(1) $f(x) = (x + 2)(x^2 + 3x - 4)$
(2) $f(x) = (x - 2)(x^2 + 3x - 4)$
(3) $f(x) = (x + 2)(x^2 + 3x + 4)$
(4) $f(x) = (x - 2)(x^2 + 3x + 4)$

4 _____

5 The cost of a pack of chewing gum in a vending machine is \$0.75. The cost of a bottle of juice in the same machine is \$1.25. Julia has \$22.00 to spend on chewing gum and bottles of juice for her team and she must buy seven packs of chewing gum. If b represents the number of bottles of juice, which inequality represents the maximum number of bottles she can buy?

(1) $0.75b + 1.25(7) \geq 22$
(2) $0.75b + 1.25(7) \leq 22$
(3) $0.75(7) + 1.25b \geq 22$
(4) $0.75(7) + 1.25b \leq 22$

5 _____

6 Which graph represents the solution of $y \leq x + 3$ and
$y \geq -2x - 2$?

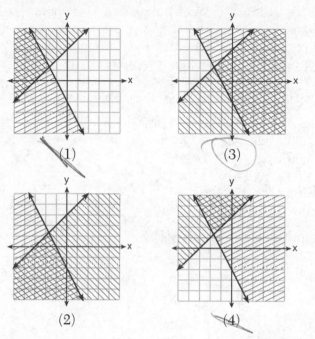

(1)   (3)

(2)   (4)

6 ___

7 The country of Benin in West Africa has a population
of 9.05 million people. The population is growing at a
rate of 3.1% each year. Which function can be used to
find the population 7 years from now?

(1) $f(t) = (9.05 \times 10^6)(1 - 0.31)^7$

(2) $f(t) = (9.05 \times 10^6)(1 + 0.31)^7$

(3) $f(t) = (9.05 \times 10^6)(1 + 0.031)^7$

(4) $f(t) = (9.05 \times 10^6)(1 - 0.031)^7$

7 ___

8 A typical cell phone plan has a fixed base fee that includes a certain amount of data and an overage charge for data use beyond the plan. A cell phone plan charges a base fee of $62 and an overage charge of $30 per gigabyte of data that exceed 2 gigabytes. If C represents the cost and g represents the total number of gigabytes of data, which equation could represent this plan when more than 2 gigabytes are used?

(1) $C = 30 + 62(2 - g)$
(2) $C = 30 + 62(g - 2)$
(3) $C = 62 + 30(2 - g)$
(4) $C = 62 + 30(g - 2)$

8 ____

9 Four expressions are shown below.

  I. $2(2x^2 - 2x - 60)$
 II. $4(x^2 - x - 30)$
III. $4(x + 6)(x - 5)$
 IV. $4x(x - 1) - 120$

The expression $4x^2 - 4x - 120$ is equivalent to

(1) I and II, only
(2) II and IV, only
(3) I, II, and IV
(4) II, III, and IV

9 ____

10 Last week, a candle store received $355.60 for selling 20 candles. Small candles sell for $10.98 and large candles sell for $27.98. How many large candles did the store sell?

(1) 6
(2) 8
(3) 10
(4) 12

10 ____

11 Which representations are functions?

I   III

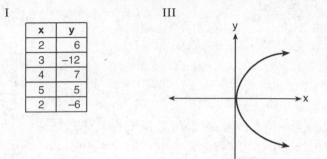

x	y
2	6
3	-12
4	7
5	5
2	-6

II $\{(1, 1), (2, 1), (3, 2), (4, 3), (5, 5), (6, 8), (7, 13)\}$ IV $y = 2x + 1$

(1) I and II   (3) II, only

(2) II and IV   (4) IV, only   11 __2__

12 If $f(x) = \dfrac{\sqrt{2x + 3}}{6x - 5}$, then $f\left(\dfrac{1}{2}\right) =$

(1) 1   (3) 1

(2) −2   (4) $-\dfrac{13}{3}$ 12 __3__

13 The zeros of the function $f(x) = 3x^2 - 3x - 6$ are

(1) −1 and −2   (3) 1 and 2

(2) 1 and −2   (4) −1 and 2   13 __4__

14 Which recursively defined function has a first term equal to 10 and a common difference of 4?

(1) $f(1) = 10$
    $f(x) = f(x - 1) + 4$

(3) $f(1) = 10$
    $f(x) = 4f(x - 1)$

(2) $f(1) = 4$
    $f(x) = f(x - 1) + 10$

(4) $f(1) = 4$
    $f(x) = 10f(x - 1)$

14 ___1___

15 Firing a piece of pottery in a kiln takes place at different temperatures for different amounts of time. The graph below shows the temperatures in a kiln while firing a piece of pottery after the kiln is preheated to 200°F.

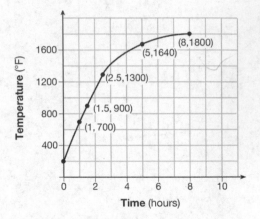

During which time interval did the temperature in the kiln show the greatest average rate of change?

(1) 0 to 1 hour

(3) 2.5 hours to 5 hours

(2) 1 hour to 1.5 hours

(4) 5 hours to 8 hours

15 ___1___

16 Which graph represents $f(x) = \begin{cases} |x| & x < 1 \\ \sqrt{x} & x \geq 1 \end{cases}$?

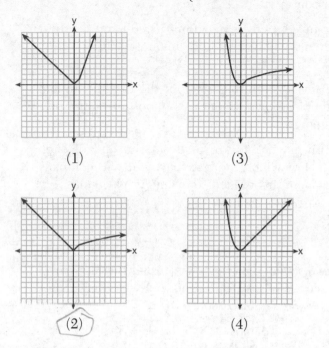

(1)   (3)

(2)   (4)

16 2

17 If $f(x) = x^2 - 2x - 8$ and $g(x) = \dfrac{1}{4}x - 1$, for which values
    of x is $f(x) = g(x)$?

(1) −1.75 and −1.438   (3) −1.438 and 0
(2) −1.75 and 4   (4) 4 and 0

17 ___ 2

18 Alicia has invented a new app for smart phones that two companies are interested in purchasing for a 2-year contract.

Company *A* is offering her $10,000 for the first month and will increase the amount each month by $5000.

Company *B* is offering $500 for the first month and will double their payment each month from the previous month.

Monthly payments are made at the end of each month. For which monthly payment will company *B*'s payment first exceed company *A*'s payment?

(1) 6   (3) 8

(2) 7   (4) 9   18 ___

19 The two sets of data below represent the number of runs scored by two different youth baseball teams over the course of a season.

Team *A*: 4, 8, 5, 12, 3, 9, 5, 2
Team *B*: 5, 9, 11, 4, 6, 11, 2, 7

Which set of statements about the mean and standard deviation is true?

(1) mean *A* < mean *B*
    standard deviation *A* > standard deviation *B*

(2) mean *A* > mean *B*
    standard deviation *A* < standard deviation *B*

(3) mean *A* < mean *B*
    standard deviation *A* < standard deviation *B*

(4) mean *A* > mean *B*
    standard deviation *A* > standard deviation *B*   19 ___

20 If Lylah completes the square for $f(x) = x^2 - 12x + 7$ in order to find the minimum, she must write $f(x)$ in the general form $f(x) = (x - a)^2 + b$. What is the value of a for $f(x)$?

(1) 6   (3) 12

(2) –6   (4) –12   20 __1__

21 Given the following quadratic functions:

$$g(x) = -x^2 - x + 6$$

and

x	–3	–2	–1	0	1	2	3	4	5
n(x)	–7	0	5	8	9	8	5	0	–7

Which statement about these functions is true?

(1) Over the interval $-1 \le x \le 1$, the average rate of change for $n(x)$ is less than that for $g(x)$.

(2) The y-intercept of $g(x)$ is greater than the y-intercept for $n(x)$.

(3) The function $g(x)$ has a greater maximum value than $n(x)$.

(4) The sum of the roots of $n(x) = 0$ is greater than the sum of the roots of $g(x) = 0$.   21

22 For which value of P and W is $P + W$ a rational number?

(1) $P = \dfrac{1}{\sqrt{3}}$ and $W = \dfrac{1}{\sqrt{6}}$

(2) $P = \dfrac{1}{\sqrt{4}}$ and $W = \dfrac{1}{\sqrt{9}}$

(3) $P = \dfrac{1}{\sqrt{6}}$ and $W = \dfrac{1}{\sqrt{10}}$

(4) $P = \dfrac{1}{\sqrt{25}}$ and $W = \dfrac{1}{\sqrt{2}}$

22 __2__

23 The solution of the equation $(x + 3)^2 = 7$ is

(1) $3 \pm \sqrt{7}$ (3) $-3 \pm \sqrt{7}$

(2) $7 \pm \sqrt{3}$ (4) $-7 \pm \sqrt{3}$

23 __3__

24 Which trinomial is equivalent to $3(x - 2)^2 - 2(x - 1)$?

(1) $3x^2 - 2x - 10$ (3) $3x^2 - 14x + 10$

(2) $3x^2 - 2x - 14$ (4) $3x^2 - 14x + 14$

24 __4__

PART II

Answer all 8 questions in this part. Each correct answer will receive 2 credits. Clearly indicate the necessary steps, including appropriate formula substitutions, diagrams, graphs, charts, etc. For all questions in this part, a correct numerical answer with no work shown will receive only 1 credit.   [16 credits]

25 Each day Toni records the height of a plant for her science lab. Her data are shown in the table below.

Day (n)	1	2	3	4	5
Height (cm)	3.0	4.5	6.0	7.5	9.0

The plant continues to grow at a constant daily rate. Write an equation to represent $h(n)$, the height of the plant on the nth day.

26 On the set of axes below, graph the inequality
$2x + y > 1$.

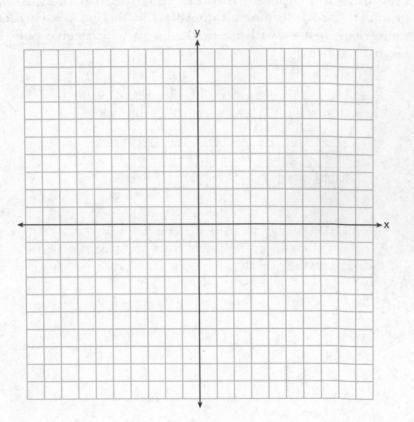

27 Rachel and Marc were given the information shown below about the bacteria growing in a Petri dish in their biology class.

Number of Hours, x	1	2	3	4	5	6	7	8	9	10
Number of Bacteria, $B(x)$	220	280	350	440	550	690	860	1070	1340	1680

Rachel wants to model this information with a linear function. Marc wants to use an exponential function. Which model is the better choice? Explain why you chose this model.

28 A driver leaves home for a business trip and drives at a constant speed of 60 miles per hour for 2 hours. Her car gets a flat tire, and she spends 30 minutes changing the tire. She resumes driving and drives at 30 miles per hour for the remaining one hour until she reaches her destination.

On the set of axes below, draw a graph that models the driver's distance from home.

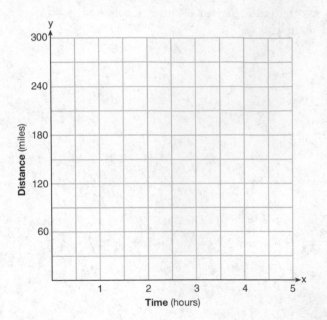

29 How many real solutions does the equation
$x^2 - 2x + 5 = 0$ have? Justify your answer.

30 The number of carbon atoms in a fossil is given by the function $y = 5100(0.95)^x$, where x represents the number of years since being discovered.

What is the percent of change each year? Explain how you arrived at your answer.

31 A toy rocket is launched from the ground straight upward. The height of the rocket above the ground, in feet, is given by the equation $h(t) = -16t^2 + 64t$, where t is the time in seconds.

Determine the domain for this function in the given context. Explain your reasoning.

32 Jackson is starting an exercise program. The first day he will spend 30 minutes on a treadmill. He will increase his time on the treadmill by 2 minutes each day. Write an equation for $T(d)$, the time, in minutes, on the treadmill on day d.

Find $T(6)$, the minutes he will spend on the treadmill on day 6.

PART III

Answer all 4 questions in this part. Each correct answer will receive 4 credits. Clearly indicate the necessary steps, including appropriate formula substitutions, diagrams, graphs, charts, etc. For all questions in this part, a correct numerical answer with no work shown will receive only 1 credit.   [16 credits]

33 Graph $f(x) = x^2$ and $g(x) = 2^x$ for $x \geq 0$ on the set of axes below.

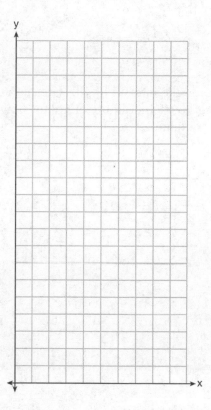

State which function, $f(x)$ or $g(x)$, has a greater value when $x = 20$. Justify your reasoning.

34 Solve for x algebraically: $7x - 3(4x - 8) \leq 6x + 12 - 9x$

If x is a number in the interval $[4, 8]$, state all integers that satisfy the given inequality. Explain how you determined these values.

35 The volume of a large can of tuna fish can be calculated using the formula $V = \pi r^2 h$. Write an equation to find the radius, r, in terms of V and h.

Determine the diameter, to the *nearest inch*, of a large can of tuna fish that has a volume of 66 cubic inches and a height of 3.3 inches.

36 The table below shows the attendance at a museum in select years from 2007 to 2013.

Attendance at Museum

Year	2007	2008	2009	2011	2013
Attendance (millions)	8.3	8.5	8.5	8.8	9.3

State the linear regression equation represented by the data table when $x = 0$ is used to represent the year 2007 and y is used to represent the attendance. Round all values to the *nearest hundredth*.

State the correlation coefficient to the *nearest hundredth* and determine whether the data suggest a strong or weak association.

PART IV

Answer the question in this part. A correct answer will receive 6 credits. Clearly indicate the necessary steps, including appropriate formula substitutions, diagrams, graphs, charts, etc. A correct numerical answer with no work shown will receive only 1 credit. [6 credits]

37 A rectangular picture measures 6 inches by 8 inches. Simon wants to build a wooden frame for the picture so that the framed picture takes up a maximum area of 100 square inches on his wall. The pieces of wood that he uses to build the frame all have the same width.

Write an equation or inequality that could be used to determine the maximum width of the pieces of wood for the frame Simon could create.

Explain how your equation or inequality models the situation.

Solve the equation or inequality to determine the maximum width of the pieces of wood used for the frame to the *nearest tenth of an inch*.

Answers
August 2015
Algebra I

Answer Key

PART I

1. (2)	**5.** (4)	**9.** (3)	**13.** (4)	**17.** (2)	**21.** (4)
2. (3)	**6.** (3)	**10.** (2)	**14.** (1)	**18.** (3)	**22.** (2)
3. (4)	**7.** (3)	**11.** (2)	**15.** (1)	**19.** (1)	**23.** (3)
4. (1)	**8.** (4)	**12.** (3)	**16.** (2)	**20.** (1)	**24.** (4)

PART II

25. $h(n) = 3 + (n-1)1.5$

26. Dotted line of $y = -2x + 1$ with shading above the line

27. Exponential model is more appropriate

28. A correct graph is drawn (see page 231)

29. No real solutions

30. 5%

31. Rocket lands after 4 seconds, domain is $0 \le t \le 4$

32. $T(d) = 30 + 2(d-1)$, $T(6) = 40$

PART III

33. $g(20) > f(20)$

34. $x \ge 6, 7, 8$

35. $r = \sqrt{\dfrac{V}{\pi h}}$, 5 inches

36. $y = 0.16x + 8.27$, $r = 0.97$, strong correlation since r is close to 1

PART IV

37. $(2x + 6)(2x + 8) \le 100$, $x = 1.5$ in

In Parts II–IV, you are required to show how you arrived at your answers. For sample methods of solutions, see the *Answers Explained* section.

Answers Explained

PART I

1. The y-intercept of the line is located at $(0, b)$. If b is increased by 4, the y-coordinate of the y-intercept will also be increased by 4. This is a shift of 4 units up.

   If, for example, b was originally 1, the line would have a slope of –2 and a y-intercept of $(0, 1)$. If the b-value was increased by 4, the new value of b would be 5. The new line would have a slope of –2 and a y-intercept of $(0, 5)$, which is a shift up by 4 units.

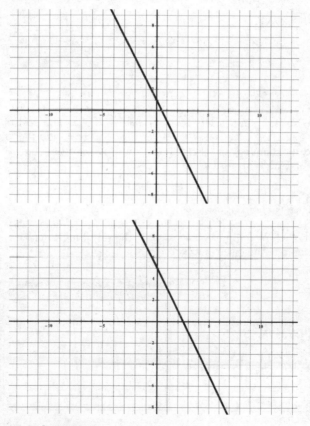

   The correct choice is **(2)**.

2. Rowan's graph can be shown by the equation $y = 50 + 5x$, while Jonah's graph can be shown by the equation $y = 10 + 15x$. Jonah's line has a slope of 15, while Rowan's line has a slope of 5. Since 15 is greater than 5, Jonah's line is steeper than Rowan's. Based on the graph, at first Rowan's line lies above Jonah's. After $x = 4$ weeks, Jonah's line lies above Rowan's.

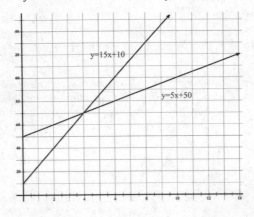

The correct choice is **(3)**.

3. The cost of a adult tickets at $3.00 each is $3.00a$. The cost of s student tickets at $1.50 each is $1.50s$. To get the total cost, add the amount collected for the adult tickets to the amount collected for the student tickets. This is the expression

$$3.00a + 1.50s.$$

The correct choice is **(4)**.

4. Graph each of the four choices using the graphing calculator, and compare them to the given graph.

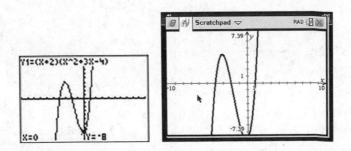

Without using the graphing calculator, the equation can be found by locating the x-intercepts of the graph. For the given graph, there are x-intercepts at $(-4, 0)$, $(-2, 0)$, and $(1, 0)$. So the function the graph is based on will have the factors $(x + 4)$, $(x + 2)$, and $(x - 1)$. In general, if there is an x-intercept at $(a, 0)$, there will be a factor of $(x - a)$.

A function that has these three factors is $f(x) = (x + 4)(x + 2)(x - 1)$. If the first and third factors are combined, the function becomes

$$f(x) = (x + 2)(x^2 + 3x - 4).$$

The correct choice is **(1)**.

5. The cost of the seven packs of chewing gum is $0.75 \cdot 7$. The cost of b bottles of juice is $1.25b$. Together the gum and the juice must cost less than or equal to \$22.00. So the inequality is $0.75(7) + 1.25b \le 22$.

The correct choice is **(4)**.

6. To graph the inequality $y \le x + 3$, first graph $y = x + 3$ with a solid line (if the inequality contained a $<$ sign instead of a $\le$, it would require a dotted line). To decide which side of the line to shade, test to see if $(0, 0)$ makes the inequality true:

$$0 \le 0 + 3$$
$$0 \le 3$$

Since this is true, the side of the line containing $(0, 0)$ gets shaded.

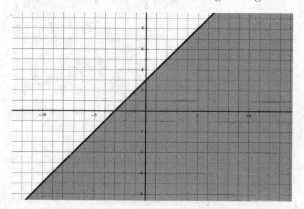

For the inequality $y \ge -2x - 2$, first graph $y = -2x - 2$ with a solid line. Test to see if $(0, 0)$ makes the inequality true:

$$0 \ge -2 \cdot 0 - 2$$
$$0 \ge -2$$

Since this is true, the side of the line containing $(0, 0)$ gets shaded.

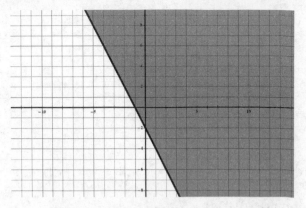

When both of these graphs are put onto the same set of axes, the graph looks like the following.

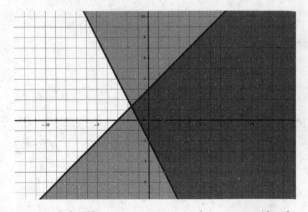

Another way to solve this question is to pick a point in the double-shaded region for each of the answer choices to see if it makes both equations true.

Testing choice (1):
$(-3, 1)$ is a point in the double-shaded region.

$$1 \le -3 + 3$$
$$1 \le 0$$

This is not true, so choice (1) can be eliminated.

Testing choice (2):
$(-2, -1)$ is a point in the double-shaded region.

$$-1 \leq -2 + 3$$
$$-1 \leq 1$$

This is true for the first inequality. Test the second inequality

$$-1 \geq -2 \cdot -2 - 2$$
$$-1 \geq 4 - 2$$
$$-1 \geq 2$$

This is not true, so choice (2) can be eliminated.

Testing choice (3):
$(2, 1)$ is a point in the double-shaded region.

$$1 \leq 2 + 3$$
$$1 \leq 5$$

This is true for the first inequality. Test for the second inequality.

$$1 \geq -2 \cdot 2 - 2$$
$$1 \geq -4 - 2$$
$$1 \geq -6$$

This is also true, so choice (3) works.

Yet another way to answer this question is to produce the graph on a graphing calculator.

The TI-Nspire allows you to graph inequalities like this. When inputting the function, delete the = and replace with the appropriate sign.

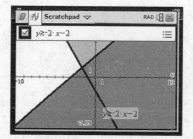

For the TI-84, you can graph the lines, but you have to determine which side of the lines to shade by testing a point like $(0, 0)$. Then the shading can be set up in the Y= menu. Move the cursor to the \ symbol to the left of Y1, and press [ENTER] three times to change the symbol to shade below the line. Move the cursor to the \ symbol to the left of Y2, and press [ENTER] two times to change the symbol to shade above the line.

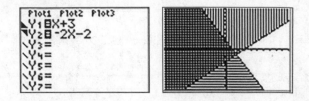

The correct choice is (**3**).

7. An exponential function based on a real-world scenario often has the form $f(x) = a(1 + r)^x$, where a is the starting value and r is the growth rate. Note that r can be negative if the scenario is based on exponential decay. For this example, the a-value is 9.05 million which, in scientific notation, is 9.05×10^6. The r-value is 0.031 for 3.1%. (It is not 0.31 which would represent 31%.) Since the question asks for an expression for seven years from now, the value of x should be 7. The function is

$$f(t) = (9.05 \times 10^6)(1 + 0.031)^7$$

The correct choice is (**3**).

8. One way to solve this question is to calculate the cost for a specific number of gigabytes, for example 10. If 10 gigabytes are used, the cost will be $62 plus $30 · 8 or $62 + $240 = $302. From the answer choices, choice (4) becomes $C = 62 + 30(10 - 2) = 62 + 30 \cdot 8$, which is the same calculation needed to find the cost for 10 gigabytes of data.

The correct choice is (**4**).

9. Multiply each of the four expressions to see which simplify to

$$4x^2 - 4x - 120$$

Testing I:
By the distributive property,

$2(2x^2 - 2x - 60) = 4x^2 - 4x - 120$

This works.

Testing II:
By the distributive property,

$4(x^2 - x - 30) = 4x^2 - 4x - 120$

This works.

Testing III:
Multiply the two binomials using the FOIL process (First, Outer, Inner, Last)

$4(x + 6)(x - 5)$

$4(x \cdot x + x(-5) + 6(x) + 6(-5))$
$4(x^2 - 5x + 6x - 30)$
$4(x^2 + x - 30)$

By the distributive property,

$4(x^2 + x - 30) = 4x^2 + 4x - 120$

This does not work because it has $+ 4x$ instead of $- 4x$.

Testing IV:
By the distributive property,

$4x(x - 1) - 120$
$4x(x) - (4x)(1) - 120$
$4x^2 - 4x - 120$
This works.

The correct choice is **(3)**.

10. Set up a system of equations. Let x be the number of small candles sold. Let y be the number of large candles sold.

Since 20 candles are sold, one equation is $x + y = 20$.

The cost of x small candles is $10.98x$. The cost of y large candles is $27.98y$. Since the total cost is known to be $355.60, the other equation is $10.98x + 27.98y = 355.60$.

x	$+$	y	$=$	20
$10.98x$	$+$	$27.98y$	$=$	355.60

Multiply both sides of the top equation by -10.98.

$-10.98(x$	$+$	$y)$	$=$	$-10.98(20)$
$10.98x$	$+$	$27.98y$	$=$	355.60

$-10.98x$	$-$	$10.98y$	$=$	-219.60
$+$				
$10.98x$	$+$	$27.98y$	$=$	355.60

		$17y$	$=$	136
		$\dfrac{17y}{17}$	$=$	$\dfrac{136}{17}$
		y	$=$	8

Since this is a multiple-choice question, there is an alternative way to solve it that does not involve creating and solving the system of equations. The correct answer can be found by testing each of the answer choices until you find one that agrees with the given information.

Testing choice (1):
If 6 large candles were sold, $20 - 6 = 14$ small candles were sold. The price of 6 large candles and 14 small candles is $6 \cdot 27.98 + 14 \cdot 10.98 = 321.60$, which is not 355.60.

Testing choice (2):
If 8 large candles were sold, $20 - 8 = 12$ small candles were sold. The price of 8 large candles and 12 small candles is $8 \cdot 27.98 + 12 \cdot 10.98 = 355.60$.

The correct choice is (2).

11. For a set of ordered pairs to be a function, each x-coordinate must have one and only one y-coordinate.

Checking I:
There is a row on the chart with a 2 for the x and a 6 for the y and another row with a 2 for the x and a -6 for the y. These correspond to the ordered pairs $(2, 6)$ and $(2, -6)$. However, a function cannot have two ordered pairs with the same x-coordinate but with different y-coordinates.

This is not a function.

Checking II:
In this list of ordered pairs, each has a different x-coordinate. So this is a function. The ordered pairs $(1, 1)$ and $(2, 1)$ have the same y-coordinate, but that is permitted. Only if two ordered pairs have the same x-coordinate but different y-coordinates would it then not be a function.

This is a function.

Checking III:
This graph fails the vertical line test. Since it is possible to draw a vertical line that passes through two points of the graph, those two points have the same x-coordinate but different y-coordinates.

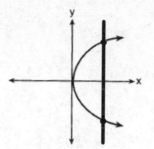

This is not a function.

Checking IV:
The graph of the solution set of the equation $y = 2x + 1$ is a line that passes the vertical line test. There is no vertical line that passes through more than one point on the line.

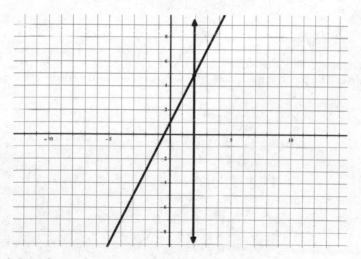

This is a function.

The correct choice is (**2**).

12. To evaluate $f\left(\dfrac{1}{2}\right)$, substitute $\dfrac{1}{2}$ for both occurrences of x in the function definition.

$$f\left(\frac{1}{2}\right) = \frac{\sqrt{2\left(\frac{1}{2}\right)+3}}{6\left(\frac{1}{2}\right)-5}$$

$$= \frac{\sqrt{1+3}}{3-5} = \frac{\sqrt{4}}{-2} = \frac{2}{-2} = -1$$

The correct choice is **(3)**.

13. The zeros of a function are the input values that make the function evaluate to zero. To find the zeros of this function, find the solutions to the equation $0 = 3x^2 - 3x - 6$. This equation is most quickly solved by factoring.

$$0 = 3x^2 - 3x - 6$$
$$0 = 3(x^2 - x - 2)$$
$$0 = 3(x - 2)(x + 1)$$
$$x - 2 = 0 \text{ or } x + 1 = 0$$
$$x = 2 \text{ or } x = -1$$

Since this is a multiple-choice question, there is a way to solve it without factoring. The numbers from the four answer choices, 1, 2, −1, and −2, can all be put into the function to see which of them evaluates to zero.

$$f(1) = 3 \cdot 1^2 - 3 \cdot 1 - 6$$
$$= 3 - 3 - 6$$
$$= -6$$

$$f(2) = 3 \cdot 2^2 - 3 \cdot 2 - 6$$
$$= 12 - 6 - 6$$
$$= 0$$

$$f(-1) = 3(-1)^2 - 3(-1) - 6$$
$$= 3 + 3 - 6$$
$$= 0$$

$$f(-2) = 3(-2)^2 - 3(-2) - 6$$
$$= 12 + 6 - 6$$
$$= 12$$

Since $f(2) = 0$ and $f(-1) = 0$, the zeros of the function are 2 and -1.

The correct choice is **(4)**.

14. A sequence with a first term equal to 10 and a common difference of 4 has the numbers 10, 14, 18, 22, ... The recursively defined function must have $f(1) = 10$ and $f(2) = 14$. Only choices (1) and (3) have $f(1) = 10$. Calculate $f(2)$ for those two choices.

Testing choice (1):

$$\begin{aligned} f(2) &= f(2-1) + 4 \\ &= f(1) + 4 \\ &= 10 + 4 \\ &= 14 \end{aligned}$$

This works.

Testing choice (3):

$$\begin{aligned} f(2) &= 4f(2-1) \\ &= 4f(1) \\ &= 4(10) \\ &= 40 \end{aligned}$$

This does not work.

The correct choice is **(1)**.

15. The average rate of change is calculated by dividing the change in temperature by the change in time.

Testing choice (1):

$$\begin{aligned} &\frac{700-200}{1-0} \\ &= \frac{500}{1} \\ &= 500 \end{aligned}$$

Testing choice (2):

$$\begin{aligned} &\frac{900-700}{1.5-1} \\ &= \frac{200}{0.5} \\ &= 400 \end{aligned}$$

Testing choice (3):

$$\frac{1640-1300}{5-2.5}$$

$$=\frac{340}{2.5}$$

$$=136$$

Testing choice (4):

$$\frac{1800-1640}{8-5}$$

$$=\frac{160}{3}$$

$$\approx 53.33$$

This question can also be answered without doing any calculations since the average rate of change is related to the slope of the line segments joining the endpoints of the interval. Since the line segment connecting $(0, 200)$ to $(1, 700)$ is the steepest of the four intervals, the rate of change on that interval is the greatest.

The correct choice is **(1)**.

16. A piecewise function behaves differently for different input values. In this example, the graph should look like $y = |x|$ for x-values less than 1 and $y = \sqrt{x}$ for x-values greater than or equal to 1.

The graph of $y = |x|$ looks like this:

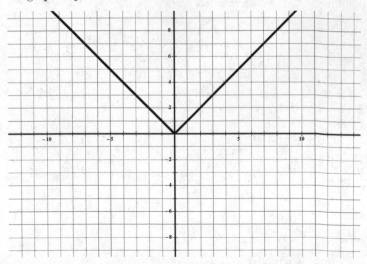

The graph of $y = \sqrt{x}$ looks like this:

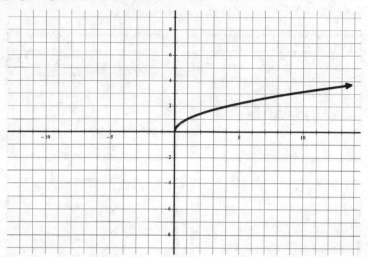

Taking the portion of $y = |x|$ to the left of $x = 1$ and the portion of $y = \sqrt{x}$ to the right of $x = 1$ looks like this:

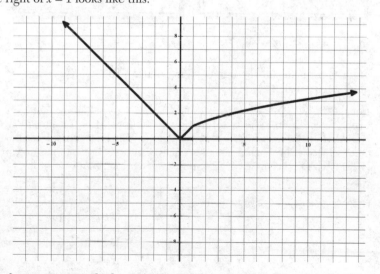

On the TI-Nspire calculator, piecewise functions can be graphed. From the Graph Scratchpad in the entry line, press the [TEMPLATE] key next to the [9]. Select the piecewise template on the top row, third from the right.

In the "Number of function pieces" field, put 2. To enter the function pieces with the conditions, press [ctrl] and [=] to select the appropriate symbol. The absolute value bars can also be found in the template menu.

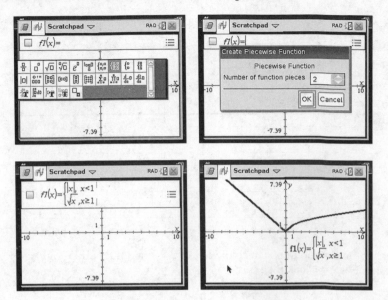

The correct choice is (2).

17. The x-coordinates of the intersection points of the two functions are the numbers for which $f(x) = g(x)$.

For the TI-84:
Press Y= and enter the two function definitions after Y1 and Y2. Press [GRAPH] [2ND] [TRACE] [5] [ENTER] [ENTER] to select the two curves. Move the cursor near one of the intersection points, and press [ENTER]. Do the same for the other intersection points.

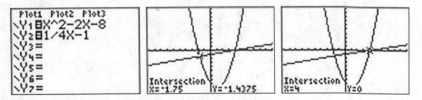

For the TI-Nspire:
From the Graph Scratchpad, enter the two functions into the entry line
and press [ENTER]. Press [MENU] [6] [4]. Move the cursor to the left
of an intersection point, and press [CLICK]. Move the cursor to the right
of the same intersection point, and press [CLICK] again.

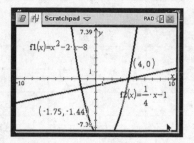

Since this is a multiple-choice question, it can also be solved by substituting the numbers from each choice into both functions to see if they
evaluate to the same number.

Testing choice (2):

$$f(-1.75) = (-1.75)^2 - 2(-1.75) - 8$$
$$= -1.4375$$
$$g(-1.75) = \frac{1}{4}(-1.75) - 1$$
$$= -1.4375$$

$$f(4) = 4^2 - 2(4) - 8$$
$$= 0$$
$$g(4) = \frac{1}{4}(4) - 1$$
$$= 0$$

Only for choice (2) is it true that $f(x) = g(x)$ for both numbers

The correct choice is **(2)**.

18. The quickest way to solve this question is to create a chart:

Month	Company A	Company B
1	10,000	500
2	15,000	1000
3	20,000	2000
4	25,000	4000
5	30,000	8000
6	35,000	16,000
7	40,000	32,000
8	45,000	64,000

Month 8 is the first time that company B's payment exceeds company A's payment.

The correct choice is **(3)**.

19. Use the graphing calculator. The mean for team A is 6, and the team's standard deviation is approximately 3.16. The mean for team B is 6.85, and the team's standard deviation is approximately 3.06. So team A has a mean less than the mean of team B. Team A has a standard deviation greater than the standard deviation of team B.

For the TI-84:
Press [STAT] [1] to enter the data. Go to the home page. Then press [STAT] [RIGHT] [1] to see the statistics.

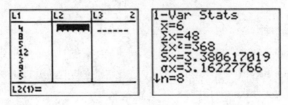

For the TI-Nspire:

From the home screen, select "Add Lists & Spreadsheet to: New Document." Enter the data into the A column starting at A1. Press [MENU] [4] [1] [1]. Enter 1 into the "Num of Lists" field. Enter a[ ] into the X1 List field. Press the "OK" button.

The correct choice is (1).

20. To complete the square for quadratic function $f(x) = x^2 + bx + c$, first rewrite it as

$$f(x) = x^2 + bx + \left(\frac{b}{2}\right)^2 - \left(\frac{b}{2}\right)^2 + c$$

and combine the last two terms.

$$f(x) = x^2 - 12x + 6^2 - 6^2 + 7$$
$$f(x) = x^2 - 12x + 36 - 36 + 7$$
$$f(x) = x^2 - 12x + 36 - 29$$

Then factor the first three terms using the perfect square trinomial pattern.

$$f(x) = (x - 6)^2 - 29$$

The a-value for this equation is 6. Had the question said $f(x) = (x + a)^2 + b$, then the a-value would have been -6.

The correct choice is (1).

21. Test each choice to determine the answer.

Testing choice (1):

For $g(x)$, the average rate of change between -1 and 1 is the change in the function $g(1) - g(-1)$ divided by the change in the x-value $1 - (-1)$.

$$\frac{g(1)-g(-1)}{1-(-1)} = \frac{(-1^2-1+6)-[-(-1)^2-(-1)+6]}{2}$$

$$= \frac{4-6}{2}$$

$$= \frac{-2}{2}$$

$$= -1$$

For $n(x)$, the average rate of change between -1 and 1 is the following:

$$\frac{n(1)-n(-1)}{1-(-1)} = \frac{9-5}{2}$$

$$= \frac{4}{2}$$

$$= 2$$

The average rate of change for $n(x)$ is greater than the average rate of change for $g(x)$ between -1 and 1, so choice (1) is not correct.

Testing choice (2):
The y-intercept of $g(x)$ is $g(0) = -0^2 - 0 + 6 = 6$. The y-intercept of $n(x)$ is $n(0) = 8$. The y-intercept of $n(x)$ is greater than the y-intercept of $g(x)$, so choice (2) is not correct.

Testing choice (3):
The maximum value of $n(x)$ is 9 at $x = 1$. The maximum value of $g(x)$ can be found with a graphing calculator. Graph the function $f(x) = -x^2 - x + 6$. Then for the TI-84, press [2ND] [TRACE] [4] to find the maximum point. For the TI-Nspire press [MENU] [6] [3] to find the maximum point.

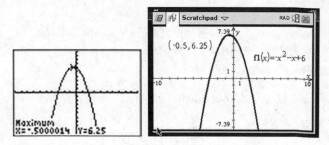

Without the graphing calculator, the maximum point of a quadratic can be calculated. The x-coordinate is $x = \dfrac{-b}{2a} = \dfrac{-(-1)}{2(-1)} = -0.5$. The y-coordinate is $g(-0.5) = 6.25$.

The maximum point is at $(-\frac{1}{2}, 6.25)$. So the maximum value is 6.25, which is less than 9. The maximum value of $n(x)$ is greater than the maximum value of $g(x)$, so choice (3) is not correct.

Testing choice (4):
The roots of a function are the input values that make that function evaluate to zero. For $n(x)$ since both $n(-2)$ and $n(4)$ equal 0, the roots of $n(x)$ are –2 and 4. For $g(x)$, the roots can be found by solving the equation $0 = -x^2 - x + 6$.

$$0 = -x^2 - x + 6$$
$$0 = -1(x^2 + x - 6)$$
$$0 = -1(x + 3)(x - 2)$$
$$x + 3 = 0 \text{ or } x - 2 = 0$$
$$x = -3 \text{ or } x = 2$$

The roots of $g(x)$ are –3 and 2.

The sum of the roots of $n(x)$ is $-2 + 4 = 2$. The sum of the roots of $g(x)$ is $-3 + 2 = -1$. So the sum of the roots of $n(x)$ is greater than the sum of the roots of $g(x)$.

Another way to test the choices is to make the graphs of $g(x)$ and $n(x)$ on the same set of axes.

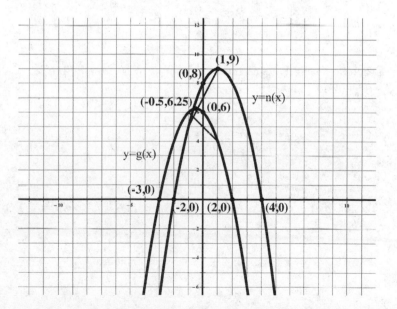

The y-intercept of $g(x)$ is less than the y-intercept of $n(x)$. The maximum value of $g(x)$ is less than the maximum value of $n(x)$. The roots correspond to the x-intercepts. The rate of change between $x = -1$ and $x = 1$ can be compared by looking at the slopes of the line segments between $x = -1$ and $x = 1$ for both parabolas. Since the line segment between $x = -1$ and $x = 1$ for $n(x)$ goes up from left to right, the slope or rate of change is positive. Since the line segment between $x = -1$ and $x = 1$ for $g(x)$ goes down from left to right, the slope or rate of change is negative. A positive rate of change is greater than any negative rate of change.

The correct choice is **(4)**.

22. The sum of two rational numbers is always rational. Since 4 and 9 are both perfect squares, $\dfrac{1}{\sqrt{4}} = \dfrac{1}{2}$ and $\dfrac{1}{\sqrt{9}} = \dfrac{1}{3}$, which are both rational numbers. So the sum $\dfrac{1}{2} + \dfrac{1}{3} = \dfrac{5}{6}$ is also rational.

The correct choice is **(2)**.

23. To solve an equation in this form, first eliminate the exponent by taking the square root of both sides of the equation:

$$(x+3)^2 = 7$$
$$\sqrt{(x+3)^2} = \pm\sqrt{7}$$
$$x+3 = \pm\sqrt{7}$$

Then eliminate the constant, 3, by subtracting it from both sides of the equation:

$$x+3 = \pm\sqrt{7}$$
$$\underline{-3 = -3\phantom{\pm\sqrt{7}}}$$
$$x = -3 \pm\sqrt{7}$$

The correct choice is **(3)**.

24. To simplify the given expression, first expand the $(x-2)^2$ with the FOIL process:

$$3(x-2)(x-2) - 2(x-1)$$
$$3(x^2 - 2x - 2x + 4) - 2(x-1)$$
$$3(x^2 - 4x + 4) - 2(x-1)$$

Now, distribute the 3 through the parentheses on the left:

$$3x^2 - 12x + 12 - 2(x-1)$$

Now, distribute the −2 through the parentheses on the right:

$$3x^2 - 12x + 12 - 2x + 2$$

Notice that the last term is +2, not −2, because $(-2)(-1) = +2$.

Finally, combine like terms:

$$3x^2 - 12x - 2x + 12 + 2$$
$$3x^2 - 14x + 14$$

An alternative way to compare the given expression to each of the answer choices is to compare the graph of $y = 3(x-2)^2 - 2(x-1)$ to the graphs of $y = 3x^2 - 2x - 10$, $y = 3x^2 - 2x - 14$, $y = 3x^2 - 14x + 10$, and $y = 3x^2 - 14x + 14$.

Since the graph of $y = 3(x-2)^2 - 2(x-1)$ is the same as the graph of $y = 3x^2 - 14x + 14$, those expressions are equivalent.

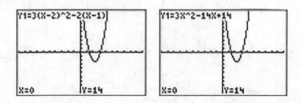

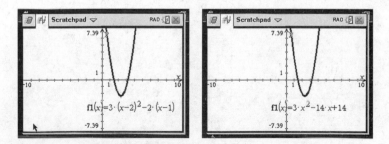

The correct choice is (4).

PART II

25. Since this is an arithmetic sequence, the formula for arithmetic sequences given in the reference sheet can be used: $a_n = a_1 + (n-1)d$, where d is the difference between two consecutive terms. This example uses function notation. So the formula is $h(n) = h(1) + (n-1)d$, where $h(1) = 3.0$ and $d = 1.5$.

One valid solution is $h(n) = 3.0 + (n-1)1.5$.

26. The graph of a linear inequality is the shading of all the points on one side of a boundary line. To graph the boundary line, change the $>$ into an $=$ and rewrite in slope-intercept form, $y = mx + b$:

$$2x + y = 1$$
$$-2x = -2x$$
$$\overline{\quad\quad\quad\quad\quad}$$
$$y = -2x + 1$$

Graph this boundary line as a dotted line since the points on the line are not part of the solution. If the inequality symbol had been a $\geq$ symbol, then the boundary line would need to be graphed as a solid line.

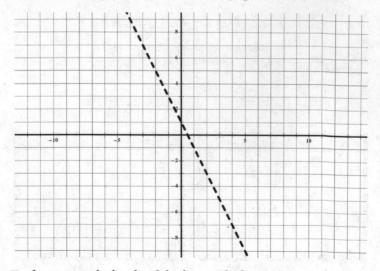

To determine which side of the line to shade, test to see whether or not $(0, 0)$ is part of the solution set. Do this by substituting 0 for both x and y into the original inequality and see if this point makes the inequality true.

$$2(0) + 0 > 1$$
$$0 > 1$$

Since this is not true, the side of the line that does not contain $(0, 0)$ should be shaded.

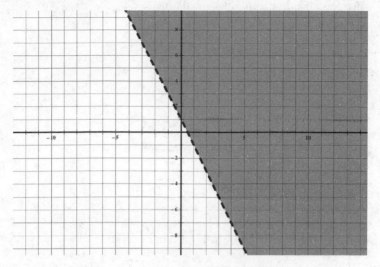

On the TI-Nspire, the inequality can be graphed by rewriting it into slope-intercept form and replacing the = with a >.

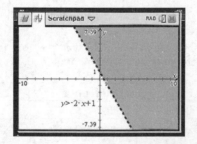

On the TI-84, you have to decide which side to shade by testing $(0, 0)$ and then setting the appropriate shading on the Y= menu by moving the cursor to the left of the Y1 on the \ and pressing [ENTER] twice. The TI-84 will not graph the boundary line as a dotted line.

27. If the shape of the graph of B(x) resembles a straight line, it can be modeled with a linear function. If the shape of the graph of B(x) is a curve, it can be modeled with an exponential function.

The graph of the 10 given points looks like this:

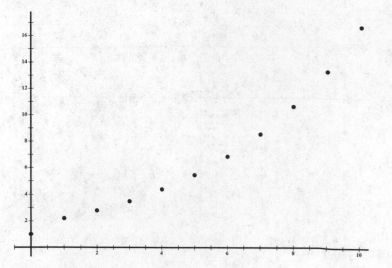

Since the graph looks more like a curve than a straight line, use an exponential model.

Even without producing the graph, it is possible to use a chart to see whether a linear function or an exponential model is more appropriate.

x	B(x)	Difference	Ratio
1	220		
2	280	$280 - 220 = 60$	$\dfrac{280}{220} \approx 1.3$
3	350	$350 - 280 = 70$	$\dfrac{350}{280} \approx 1.3$
4	440	$440 - 350 = 90$	$\dfrac{440}{350} \approx 1.3$
5	550	$550 - 440 = 110$	$\dfrac{550}{440} \approx 1.3$
6	690	$690 - 550 = 140$	$\dfrac{690}{550} \approx 1.3$

x	$B(x)$	Difference	Ratio
7	860	$860 - 690 = 170$	$\dfrac{860}{690} \approx 1.2$
8	1070	$1070 - 860 = 210$	$\dfrac{1070}{860} \approx 1.2$
9	1340	$1340 - 1070 = 270$	$\dfrac{1340}{1070} \approx 1.3$
10	1680	$1680 - 1340 = 340$	$\dfrac{1680}{1340} \approx 1.3$

If the differences in column 3 were all approximately the same number, a linear model would be more accurate. Since the ratios in column 4 are all approximately the same number, an exponential model is more accurate.

28. At the beginning of the trip, the car has not traveled any distance. This corresponds to the point (0, 0). After 2 hours, the car has traveled $2 \cdot 60 = 120$ miles. This corresponds to the point (2, 120). Since the car does not move for the next 30 minutes, it is still at 120 miles from the starting point 2.5 hours after starting. This corresponds to the point (2.5, 120). The car then travels 30 miles per hour for the next hour for a total of 30 more miles. The point corresponding to the end of the trip, then, is (3.5, 150). When these four points are joined with line segments, the graph looks like this:

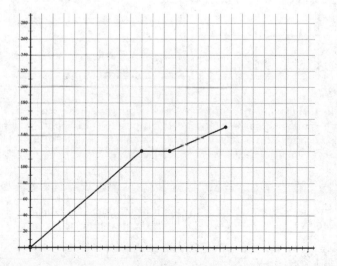

29. The solutions to a quadratic equation can be calculated with the quadratic formula found on the reference sheet, $x = \dfrac{-b \pm \sqrt{b^2 - 4ac}}{2a}$, where a, b, and c are the coefficients when the equation is in the form $ax^2 + bx + c = 0$. For this question, $a = 1$, $b = -2$, and $c = 5$.

Use the quadratic formula,

$$x = \frac{-(-2) \pm \sqrt{(-2)^2 - 4(1)(5)}}{2(1)}$$

$$= \frac{2 \pm \sqrt{4 - 20}}{2}$$

$$= \frac{2 \pm \sqrt{-16}}{2}$$

Since there is a negative number inside the square root sign, there are no real solutions to the equation.

Another way to solve this is to make a graph of $y = x^2 - 2x + 5$ and see if it has any x-intercepts. Since there are no x-intercepts, the equation $x^2 - 2x + 5 = 0$ has no real solutions.

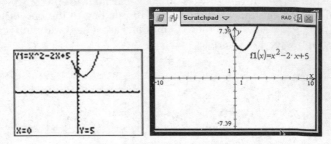

30. When an exponential equation is written in the form $y = a(1 + r)^x$, then r is the percent change each year. If r is positive, this is the growth rate. If r is negative, this is the decay rate. Since 0.95 is equal to $(1 - 0.05)$, the percent change equals -0.05, which is a 5% change.

31. The t-values for which the height $h(t)$ is measured start when the toy rocket is launched and stop when the toy rocket lands on the ground. Between those two times is the appropriate domain for this function. The toy rocket launches at $t = 0$ and lands the next time $h(t) = 0$.

Solve the equation $0 = -16t^2 + 64t$ to find when the toy rocket lands:

$$0 = -16t^2 + 64t$$
$$0 = -16t(t - 4)$$

$$-16t = 0 \text{ or } t - 4 = 0$$
$$t = 0 \text{ or } t = 4$$

So the toy rocket lands after 4 seconds, making the domain $0 \le t \le 4$.

Graphing the function on the graphing calculator also reveals that the toy rocket is launched at $t = 0$ and lands at $t = 4$.

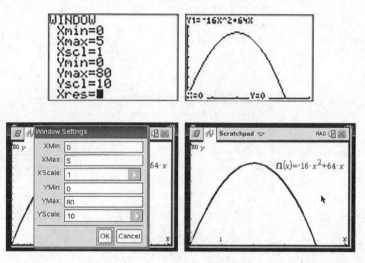

32. The number of minutes Jackson spends on the treadmill can be written as an arithmetic sequence 30, 32, 34, 36, 38, . . . where the first term is 30 and the common difference is 2.

For the equation, the formula for the nth term of an arithmetic sequence provided on the reference sheet can be used, $a_n = a_1 + (n - 1)d$, where a_1 is the first term and d is the common difference.

For this question, the function uses d instead of n for the term number.

The function is $T(d) = 30 + 2(d - 1)$, which can be simplified to $T(d) = 2d + 28$.

For the second part, substitute 6 for d into the expression:

$$T(6) = 30 + 2(6 - 1)$$
$$= 30 + 2 \cdot 5$$
$$= 40$$

PART III

33. For $x = 0$ to $x = 10$, make a chart for $f(x)$ and $g(x)$.

x	$f(x) = x^2$	$g(x) = 2^x$
0	0	1
1	1	2
2	4	4
3	9	8
4	16	16
5	25	32
6	36	64
7	49	128
8	64	256
9	81	512
10	100	1024

In order for most of the points to fit on the graph, the scale of the y-axis should be at least 30 for each box. For the graph below, each y-axis unit is 30. So the top left corner is at $(0, 600)$.

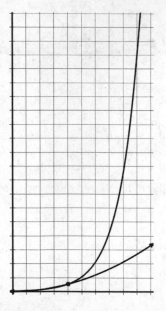

According to the graph it seems that for any x-value greater than 4 the graph of $g(x)$ is above the graph of $f(x)$. So for $x = 20$, $g(x) > f(x)$.

To be sure, you can calculate $f(20) = 20^2 = 400$ and $g(20) = 2^{20} = 1{,}048{,}576$. So $g(20)$ is much greater than $f(20)$.

34. To solve the inequality, first simplify both sides:

$$7x - 3(4x - 8) \le 6x + 12 - 9x$$
$$7x - 12x + 24 \le -3x + 12$$
$$-5x + 24 \le -3x + 12$$

Be careful in the second step to write $+24$ since $-3(-8) = +24$.

Add $3x$ to both sides of the inequality and subtract 24 from both sides of the inequality:

$$
\begin{aligned}
-5x + 24 &\le -3x + 12 \\
+3x &= +3x \\
\hline
-2x + 24 &\le 12 \\
-24 &= -24 \\
\hline
-2x &\le -12
\end{aligned}
$$

Divide both sides of the inequality by –2. When both sides of an inequality are multiplied by or divided by a negative, the direction of the inequality sign must be reversed to keep the inequality true:

$$\frac{-2x}{-2} \ge \frac{-12}{-2}$$
$$x \ge 6$$

The solution set is $x \ge 6$.

The numbers in this solution set that are also integers between 4 and 8 are just 6, 7, and 8.

35. To isolate the r-term in the equation $V = \pi r^2 h$, first treat the other variables, V and h, as if they were constants:

$$V = \pi r^2 h$$
$$\frac{V}{\pi h} = \frac{\pi r^2 h}{\pi h}$$
$$\frac{V}{\pi h} = r^2$$

Finish by taking the square root of both sides:

$$\sqrt{\frac{V}{\pi h}} = \sqrt{r^2}$$

$$\sqrt{\frac{V}{\pi h}} = r$$

Since the radius must be a positive number, the ± symbol is not necessary.

The formula for r can be used to calculate the radius of the can. Since the diameter is double the radius, multiply r by 2:

$$r = \sqrt{\frac{V}{\pi h}}$$

$$= \sqrt{\frac{66}{\pi \cdot 3.3}}$$

$$\approx 2.52$$

$$d = 2r$$
$$= 2 \cdot 2.52$$
$$= 5 \text{ inches}$$

36. The line of best fit can be found with a graphing calculator.

For the TI-84:
Turn "diagnostics" on by pressing [2ND] [0] and scroll down to DiagnosticOn and press [ENTER] [ENTER]. Press [STAT] [1]. Clear the contents of L1 and L2 by moving the cursor to the top row and pressing [CLEAR] [ENTER] for the first two columns. Input the numbers 0, 1, 2, 4, and 6 into L1. Enter 8.3, 8.5, 8.5, 8.8, and 9.3 into L2. Press [STAT] [RIGHT]. Press [4] for LinReg($ax + b$), and then press [ENTER].

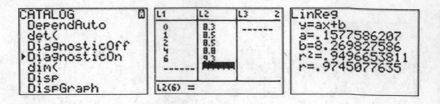

For the TI-Nspire:
From the home screen go to "Add Lists & Spreadsheet to: New Document." In cells A1 to A5, input the numbers 0, 1, 2, 4, and 6. In cells B1 to B5, enter the numbers 8.3, 8.5, 8.5, 8.8, and 9.3. Press [menu] [4] [1] [3] for Linear Regression ($mx + b$). In the X List field, enter "a[ ]" (use [ctrl] [(] to create the square brackets). In the Y List field enter "b[ ]." Press the OK button.

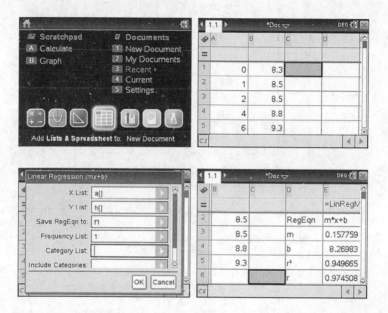

The equation with values rounded to the nearest hundredth is

$$y = 0.16x + 8.27$$

The correlation coefficient, r, rounded to the nearest hundredth is 0.97. Since this is very close to 1, there is a strong association.

PART IV

37. The diagram below shows the picture with the frame. The dimensions of the picture are 6 by 8, while the width of the frame is the unknown, denoted by x.

The width of the large rectangle is $2x + 6$, and the length of the large rectangle is $2x + 8$. Since the area of the large rectangle, which includes the frame, must be less than or equal to 100 and the area of the large rectangle is $(2x + 6)(2x + 8)$, the inequality is $(2x + 6)(2x + 8) \le 100$.

The maximum width for the frame occurs when the area is exactly 100. So the equation $(2x + 6)(2x + 8) = 100$ can be used to find the maximum width:

$$(2x \cdot 2x) + (2x \cdot 8) + (6 \cdot 2x) + (6 \cdot 8) = 100$$
$$4x^2 + 16x + 12x + 48 = 100$$
$$4x^2 + 28x + 48 = 100$$
$$-100 = -100$$

$$\overline{}$$

$$4x^2 + 28x - 52 = 0$$
$$4(x^2 + 7x - 13) = 0$$

The quadratic equation $x^2 + 7x - 13 = 0$ can be solved with the quadratic formula where $a = 1, b = 7$, and $c = -13$:

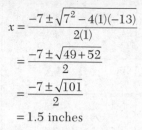

$$= 1.5 \text{ inches}$$

The other solution, $\dfrac{-7-\sqrt{101}}{2} \approx -8.5$, should be rejected since the width of the frame must be a positive number of inches.

To check that the answer is correct, see if the area of the large rectangle is close to 100. If the frame has a width of 1.5, the width of the rectangle is $2 \cdot 1.5 + 6 = 9$. Then the length of the rectangle is $2 \cdot 1.5 + 8 = 11$. The area is $9 \cdot 11 = 99$ square inches, which is close to 100 square inches.

Topic	Question Numbers	Number of Points	Your Points	Your Percentage
1. Polynomials	9, 24	2 + 2 = 4		
2. Properties of Algebra	35	4		
3. Functions	4, 11, 12	2 + 2 + 2 = 6		
4. Creating and Interpreting Equations	2, 3, 8, 15	2 + 2 + 2 + 2 = 8		
5. Inequalities	5, 6, 26, 34	2 + 2 + 2 + 4 = 10		
6. Sequences and Series	14, 18, 25, 32	2 + 2 + 2 + 2 = 8		
7. Systems of Equations	10	2		
8. Quadratic Equations and Factoring	13, 17, 20, 23, 29, 31, 37	2 + 2 + 2 + 2 + 2 + 2 + 6 = 18		
9. Regression	36	4		
10. Exponential Equations	7, 27, 30, 33	2 + 2 + 2 + 4 = 10		
11. Graphing	1, 16, 21, 28	2 + 2 + 2 + 2 = 8		
12. Statistics	19	2		
13. Number Properties	22	2		

HOW TO CONVERT YOUR RAW SCORE TO YOUR ALGEBRA I REGENTS EXAMINATION SCORE

The accompanying conversion chart must be used to determine your final score on the August 2015 Regents Examination in Algebra I. To find your final exam score, locate in the column labeled "Raw Score" the total number of points you scored out of a possible 86 points. Since partial credit is allowed in Parts II, III, and IV of the test, you may need to approximate the credit you would receive for a solution that is not completely correct. Then locate in the adjacent column to the right the scale score that corresponds to your raw score. The scale score is your final Algebra I Regents Examination score.

Regents Examination in Algebra I—August 2015
Chart for Converting Total Test Raw Scores to Final
Examination Scores (Scaled Scores)

Raw Score	Scale Score	Performance Level	Raw Score	Scale Score	Performance Level	Raw Score	Scale Score	Performance Level
86	100	5	57	75	4	28	64	2
85	99	5	56	75	4	27	63	2
84	97	5	55	75	4	26	62	2
83	96	5	54	74	4	25	61	2
82	96	5	53	73	3	24	60	2
81	93	5	52	73	3	23	59	2
80	92	5	51	73	3	22	58	2
79	91	5	50	72	3	21	57	2
78	90	5	49	72	3	20	55	2
77	89	5	48	72	3	19	54	1
76	87	5	47	72	3	18	52	1
75	86	5	46	71	3	17	51	1
74	85	5	45	71	3	16	49	1
73	84	4	44	71	3	15	47	1
72	83	4	43	71	3	14	45	1
71	83	4	42	70	3	13	43	1
70	82	4	41	70	3	12	41	1
69	81	4	40	70	3	11	38	1
68	80	4	39	69	3	10	36	1
67	80	4	38	69	3	9	33	1
66	79	4	37	69	3	8	30	1
65	78	4	36	68	3	7	27	1
64	78	4	35	68	3	6	24	1
63	77	4	34	67	3	5	21	1
62	77	4	33	67	3	4	17	1
61	76	4	32	66	3	3	13	1
60	76	4	31	66	3	2	9	1
59	76	4	30	65	3	1	5	1
58	75	4	29	64	2	0	0	1

Examination
June 2016
Algebra I

HIGH SCHOOL MATH REFERENCE SHEET

Conversions

1 inch = 2.54 centimeters

1 meter = 39.37 inches

1 mile = 5280 feet

1 mile = 1760 yards

1 mile = 1.609 kilometers

1 kilometer = 0.62 mile

1 pound = 16 ounces

1 pound = 0.454 kilogram

1 kilogram = 2.2 pounds

1 ton = 2000 pounds

1 cup = 8 fluid ounces

1 pint = 2 cups

1 quart = 2 pints

1 gallon = 4 quarts

1 gallon = 3.785 liters

1 liter = 0.264 gallon

1 liter = 1000 cubic centimeters

Formulas

Triangle	$A = \dfrac{1}{2}bh$
Parallelogram	$A = bh$
Circle	$A = \pi r^2$
Circle	$C = \pi d$ or $C = 2\pi r$

Formulas (continued)

General Prisms	$V = Bh$
Cylinder	$V = \pi r^2 h$
Sphere	$V = \dfrac{4}{3}\pi r^3$
Cone	$V = \dfrac{1}{3}\pi r^2 h$
Pyramid	$V = \dfrac{1}{3}Bh$
Pythagorean Theorem	$a^2 + b^2 = c^2$
Quadratic Formula	$x = \dfrac{-b \pm \sqrt{b^2 - 4ac}}{2a}$
Arithmetic Sequence	$a_n = a_1 + (n-1)d$
Geometric Sequence	$a_n = a_1 r^{n-1}$
Geometric Series	$S_n = \dfrac{a_1 - a_1 r^n}{1 - r}$ where $r \neq 1$
Radians	$1 \text{ radian} = \dfrac{180}{\pi} \text{ degrees}$
Degrees	$1 \text{ degree} = \dfrac{\pi}{180} \text{ radians}$
Exponential Growth/Decay	$A = A_0 e^{k(t - t_0)} + B_0$

PART I

Answer all 24 questions in this part. Each correct answer will receive 2 credits. No partial credit will be allowed. For each statement or question, write in the space provided the numeral preceding the word or expression that best completes the statement or answers the question. [48 credits]

1 The expression $x^4 - 16$ is equivalent to

(1) $(x^2 + 8)(x^2 - 8)$ (3) $(x^2 + 4)(x^2 - 4)$

(2) $(x^2 - 8)(x^2 - 8)$ (4) $(x^2 - 4)(x^2 - 4)$ 1____

2 An expression of the fifth degree is written with a leading coefficient of seven and a constant of six. Which expression is correctly written for these conditions?

(1) $6x^5 + x^4 + 7$ (3) $6x^7 - x^5 + 5$

(2) $7x^6 - 6x^4 + 5$ (4) $7x^5 + 2x^2 + 6$ 2____

3 The table below shows the year and the number of households in a building that had high-speed broadband internet access.

Number of Households	11	16	23	33	42	47
Year	2002	2003	2004	2005	2006	2007

For which interval of time was the average rate of change the *smallest*?

(1) 2002–2004   (3) 2004–2006

(2) 2003–2005   (4) 2005–2007   3____

4 The scatterplot below compares the number of bags of popcorn and the number of sodas sold at each performance of the circus over one week.

Popcorn Sales and Soda Sales

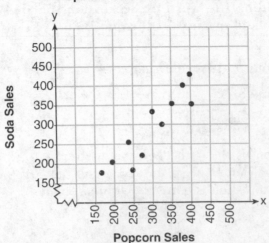

Which conclusion can be drawn from the scatterplot?

(1) There is a negative correlation between popcorn sales and soda sales.

(2) There is a positive correlation between popcorn sales and soda sales.

(3) There is no correlation between popcorn sales and soda sales.

(4) Buying popcorn causes people to buy soda.

4 _____

5 The Celluloid Cinema sold 150 tickets to a movie. Some of these were child tickets and the rest were adult tickets. A child ticket cost $7.75 and an adult ticket cost $10.25. If the cinema sold $1470 worth of tickets, which system of equations could be used to determine how many adult tickets, a, and how many child tickets, c, were sold?

(1) $a + c = 150$
$10.25a + 7.75c = 1470$

(3) $a + c = 150$
$7.75a + 10.25c = 1470$

(2) $a + c = 1470$
$10.25a + 7.75c = 150$

(4) $a + c = 1470$
$7.75a + 10.25c = 150$

5 _____

6 The tables below show the values of four different functions for given values of x.

x	f(x)
1	12
2	19
3	26
4	33

x	g(x)
1	−1
2	1
3	5
4	13

x	h(x)
1	9
2	12
3	17
4	24

x	k(x)
1	−2
2	4
3	14
4	28

Which table represents a linear function?

(1) $f(x)$

(3) $h(x)$

(2) $g(x)$

(4) $k(x)$

6 _____

7 The acidity in a swimming pool is considered normal if the average of three pH readings, p, is defined such that $7.0 < p < 7.8$. If the first two readings are 7.2 and 7.6, which value for the third reading will result in an overall rating of normal?

(1) 6.2

(3) 8.6

(2) 7.3

(4) 8.8

7 _____

8 Dan took 12.5 seconds to run the 100-meter dash. He calculated the time to be approximately

  (1) 0.2083 minute   (3) 0.2083 hour

  (2) 750 minutes   (4) 0.52083 hour   8 _____

9 When $3x + 2 \le 5(x - 4)$ is solved for x, the solution is

  (1) $x \le 3$ (3) $x \le -11$

  (2) $x \ge 3$ (4) $x \ge 11$ 9 _____

10 The expression $3(x^2 - 1) - (x^2 - 7x + 10)$ is equivalent to

  (1) $2x^2 - 7x + 7$ (3) $2x^2 - 7x + 9$

  (2) $2x^2 + 7x - 13$ (4) $2x^2 + 7x - 11$ 10 _____

11 The range of the function $f(x) = x^2 + 2x - 8$ is all real numbers

  (1) less than or equal to –9

  (2) greater than or equal to –9

  (3) less than or equal to –1

  (4) greater than or equal to –1   11 _____

12 The zeros of the function $f(x) = x^2 - 5x - 6$ are

  (1) –1 and 6   (3) 2 and –3

  (2) 1 and –6   (4) –2 and 3   12 _____

13 In a sequence, the first term is 4 and the common difference is 3. The fifth term of this sequence is

  (1) –11   (3) 16

  (2) –8   (4) 19   13 _____

14 The growth of a certain organism can be modeled by $C(t) = 10(1.029)^{24t}$, where $C(t)$ is the total number of cells after t hours. Which function is approximately equivalent to $C(t)$?

(1) $C(t) = 240(.083)^{24t}$ (3) $C(t) = 10(1.986)^t$

(2) $C(t) = 10(.083)^t$ (4) $C(t) = 240(1.986)^{\frac{t}{24}}$ 14 _____

15 A public opinion poll was taken to explore the relationship between age and support for a candidate in an election. The results of the poll are summarized in the table below.

Age	For	Against	No Opinion
21–40	30	12	8
41–60	20	40	15
Over 60	25	35	15

What percent of the 21–40 age group was for the candidate?

(1) 15   (3) 40
(2) 25   (4) 60   15 _____

16 Which equation and ordered pair represent the correct vertex form and vertex for $j(x) = x^2 - 12x + 7$?

(1) $j(x) = (x - 6)^2 + 43$, (6, 43)
(2) $j(x) = (x - 6)^2 + 43$, (−6, 43)
(3) $j(x) = (x - 6)^2 - 29$, (6, −29)
(4) $j(x) = (x - 6)^2 - 29$, (−6, −29)   16 _____

17 A student invests $500 for 3 years in a savings account that earns 4% interest per year. No further deposits or withdrawals are made during this time. Which statement does *not* yield the correct balance in the account at the end of 3 years?

(1) $500(1.04)^3$
(2) $500(1 - .04)^3$
(3) $500(1 + .04)(1 + .04)(1 + .04)$
(4) $500 + 500(.04) + 520(.04) + 540.8(.04)$

17 _____

18 The line represented by the equation $4y + 2x = 33.6$ shares a solution point with the line represented by the table below.

x	y
-5	3.2
-2	3.8
2	4.6
4	5
11	6.4

The solution for this system is

(1) $(-14.0, -1.4)$ (3) $(1.9, 4.6)$
(2) $(-6.8, 5.0)$ (4) $(6.0, 5.4)$

18 _____

19 What is the solution of the equation $2(x + 2)^2 - 4 = 28$?

(1) 6, only   (3) 2 and -6
(2) 2, only   (4) 6 and -2

19 _____

20 The dot plot shown below represents the number of pets owned by students in a class.

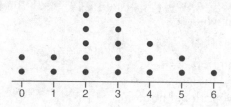

Which statement about the data is *not* true?

(1) The median is 3.
(2) The interquartile range is 2.
(3) The mean is 3.
(4) The data contain no outliers.   20 _____

21 What is the largest integer, x, for which the value of $f(x) = 5x^4 + 30x^2 + 9$ will be greater than the value of $g(x) = 3^x$?

(1) 7   (3) 9
(2) 8   (4) 10   21 _____

22 The graphs of the functions $f(x) = |x - 3| + 1$ and $g(x) = 2x + 1$ are drawn. Which statement about these functions is true?

(1) The solution to $f(x) = g(x)$ is 3.
(2) The solution to $f(x) = g(x)$ is 1.
(3) The graphs intersect when $y = 1$.
(4) The graphs intersect when $x = 3$.   22 _____

23 A store sells self-serve frozen yogurt sundaes. The function $C(w)$ represents the cost, in dollars, of a sundae weighing w ounces. An appropriate domain for the function would be

(1) integers
(2) rational numbers
(3) nonnegative integers
(4) nonnegative rational numbers

23 _____

24 Sara was asked to solve this word problem: "The product of two consecutive integers is 156. What are the integers?" What type of equation should she create to solve this problem?

(1) linear   (3) exponential
(2) quadratic   (4) absolute value

24 _____

PART II

Answer all 8 questions in this part. Each correct answer will receive 2 credits. Clearly indicate the necessary steps, including appropriate formula substitutions, diagrams, graphs, charts, etc. For all questions in this part, a correct numerical answer with no work shown will receive only 1 credit.   [16 credits]

25 Given that $f(x) = 2x + 1$, find $g(x)$ if $g(x) = 2[f(x)]^2 - 1$.

26 Determine if the product of $3\sqrt{2}$ and $8\sqrt{18}$ is rational or irrational. Explain your answer.

27 On the set of axes below, draw the graph of $y = x^2 - 4x - 1$.

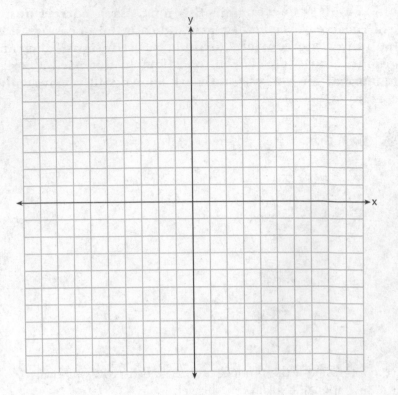

State the equation of the axis of symmetry.

28 Amy solved the equation $2x^2 + 5x - 42 = 0$. She stated that the solutions to the equation were $\frac{7}{2}$ and –6. Do you agree with Amy's solutions? Explain why or why not.

29 Sue and Kathy were doing their algebra homework. They were asked to write the equation of the line that passes through the points $(-3, 4)$ and $(6, 1)$. Sue wrote $y - 4 = -\frac{1}{3}(x + 3)$ and Kathy wrote $y = -\frac{1}{3}x + 3$. Justify why both students are correct.

30 During a recent snowstorm in Red Hook, NY, Jaime noted that there were 4 inches of snow on the ground at 3:00 p.m., and there were 6 inches of snow on the ground at 7:00 p.m.

If she were to graph these data, what does the slope of the line connecting these two points represent in the context of this problem?

31 The formula for the sum of the degree measures of the interior angles of a polygon is $S = 180(n - 2)$. Solve for n, the number of sides of the polygon, in terms of S.

32 In the diagram below, $f(x) = x^3 + 2x^2$ is graphed. Also graphed is $g(x)$, the result of a translation of $f(x)$.

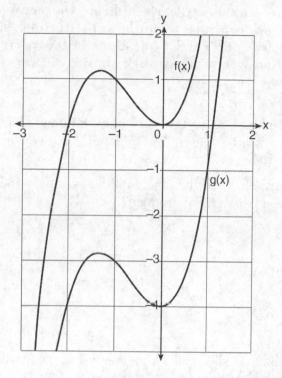

Determine an equation of $g(x)$. Explain your reasoning.

PART III

Answer all 4 questions in this part. Each correct answer will receive 4 credits. Clearly indicate the necessary steps, including appropriate formula substitutions, diagrams, graphs, charts, etc. For all questions in this part, a correct numerical answer with no work shown will receive only 1 credit. [16 credits]

33 The height, H, in feet, of an object dropped from the top of a building after t seconds is given by $H(t) = -16t^2 + 144$.

How many feet did the object fall between one and two seconds after it was dropped?

Determine, algebraically, how many seconds it will take for the object to reach the ground.

34 The sum of two numbers, x and y, is more than 8. When you double x and add it to y, the sum is less than 14.

Graph the inequalities that represent this scenario on the set of axes below.

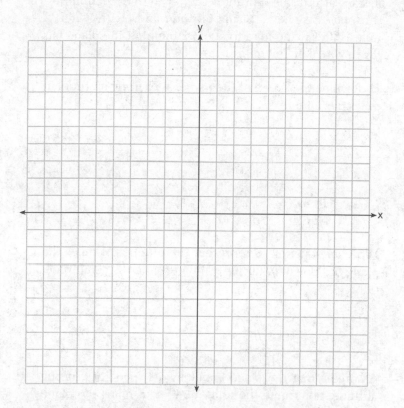

Kai says that the point (6, 2) is a solution to this system. Determine if he is correct and explain your reasoning.

35 An airplane leaves New York City and heads toward Los Angeles. As it climbs, the plane gradually increases its speed until it reaches cruising altitude, at which time it maintains a constant speed for several hours as long as it stays at cruising altitude. After flying for 32 minutes, the plane reaches cruising altitude and has flown 192 miles. After flying for a total of 92 minutes, the plane has flown a total of 762 miles.

Determine the speed of the plane, at cruising altitude, in miles per minute.

Write an equation to represent the number of miles the plane has flown, y, during x minutes at cruising altitude, only.

Assuming that the plane maintains its speed at cruising altitude, determine the total number of miles the plane has flown 2 hours into the flight.

36 On the set of axes below, graph

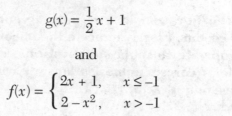

$$g(x) = \frac{1}{2}x + 1$$

and

$$f(x) = \begin{cases} 2x + 1, & x \le -1 \\ 2 - x^2, & x > -1 \end{cases}$$

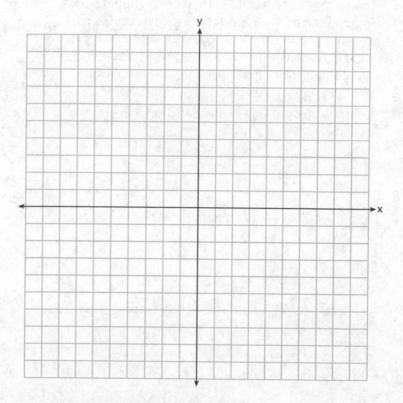

How many values of x satisfy the equation $f(x) = g(x)$?
Explain your answer, using evidence from your graphs.

PART IV

Answer the question in this part. A correct answer will receive 6 credits. Clearly indicate the necessary steps, including appropriate formula substitutions, diagrams, graphs, charts, etc. A correct numerical answer with no work shown will receive only 1 credit.   [6 credits]

37 Franco and Caryl went to a bakery to buy desserts. Franco bought 3 packages of cupcakes and 2 packages of brownies for $19. Caryl bought 2 packages of cupcakes and 4 packages of brownies for $24. Let x equal the price of one package of cupcakes and y equal the price of one package of brownies.

Write a system of equations that describes the given situation.

On the set of axes below, graph the system of equations.

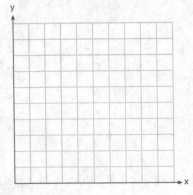

Determine the exact cost of one package of cupcakes and the exact cost of one package of brownies in dollars and cents. Justify your solution.

Answers
June 2016

Algebra I

Answer Key

PART I

1. (3)	**5.** (1)	**9.** (4)	**13.** (3)	**17.** (2)	**21.** (3)
2. (4)	**6.** (1)	**10.** (2)	**14.** (3)	**18.** (4)	**22.** (2)
3. (1)	**7.** (2)	**11.** (2)	**15.** (4)	**19.** (3)	**23.** (4)
4. (2)	**8.** (1)	**12.** (1)	**16.** (3)	**20.** (3)	**24.** (2)

PART II

25. $8x^2 + 8x + 1$

26. 144 is a rational number.

27. $x = 2$

28. She is correct.

29. They are both correct.

30. $\frac{1}{2}$ is the number of inches of snow that falls in one hour.

31. $n = \dfrac{S + 360}{180}$

32. $g(x) = f(x) - 4$

PART III

33. $48, t = 3$

34. The solution set does not contain $(6, 2)$.

35. 9.5 miles per minute, $y = 9.5x$, 1,028

36. One intersection point

PART IV

37. \$3.50 and \$4.25

In Parts II–IV, you are required to show how you arrived at your answers. For sample methods of solutions, see the *Answers Explained* section.

Answers Explained

PART I

1. The difference of perfect squares factoring pattern is $a^2 - b^2 = (a + b)(a - b)$. The expression $x^4 - 16$ can be written as $(x^2)^2 - 4^2$. So

$$(x^2)^2 - 4^2 = (x^2 + 4)(x^2 - 4)$$

Another way to find this answer is to multiply each of the four answer choices to see which one becomes $x^4 - 16$. When using FOIL, choice (3) would become $(x^2 + 4)(x^2 - 4) = x^4 - 4x^2 + 4x^2 - 16 = x^4 - 16$.

The correct choice is **(3)**.

2. The degree of a polynomial expression is the greatest exponent to which the variable is raised. A fifth-degree polynomial expression has an x^5-term as its highest term. The leading coefficient is the number that is multiplied by the variable raised to the highest exponent. If the leading term of a fifth-degree polynomial expression is 7, one of the terms of that expression will be $7x^5$. The constant in a polynomial expression is the number that is not being multiplied by any variable. When the constant is 6, there will be a +6 in the expression. Only choice (4) contains the terms $7x^5$ and a +6.

The correct choice is **(4)**.

3. The average rate of change of an interval can be calculated for this problem with the following formula.

$$\text{Average Rate of Change} = \frac{\text{Change in Households}}{\text{Change in Years}}$$

All 4 choices must be tested.

Testing choice (1):
$$\text{Average Rate of Change} = \frac{23 - 11}{2004 - 2002} = \frac{12}{2} = 6$$

Testing choice (2):
$$\text{Average Rate of Change} = \frac{33 - 16}{2005 - 2003} = \frac{17}{2} = 8.5$$

Testing choice (3):

$$\text{Average Rate of Change} = \frac{42-23}{2006-2004} = \frac{19}{2} = 9.5$$

Testing choice (4):

$$\text{Average Rate of Change} = \frac{47-33}{2007-2005} = \frac{14}{2} = 7$$

Of the four choices, choice (1) has the smallest average rate of change.

The correct choice is (**1**).

4. When the points of a scatterplot resemble a line with a positive slope, we say that there is a positive correlation between the two variables. This scatterplot does resemble a line with a positive slope.

Choice (1) is incorrect because a negative correlation exists when the scatterplot resembles a line with a negative slope.

Choice (3) is incorrect because no correlation exists when the scatterplot does not resemble a line at all.

Choice (4) is the second-best choice. It is possible that buying popcorn causes people to buy soda. However, we cannot confidently draw that conclusion from the scatterplot. Perhaps it is the soda that caused people to want popcorn or something else that causes both things to happen. All that can be said for sure is that there is a positive correlation between the popcorn sales and soda sales.

The correct choice is (**2**).

5. Let a be the number of adult tickets sold and c be the number of child tickets sold. Since there were 150 tickets sold, one of the equations must be $a + c = 150$. This eliminates choices (2) and (4). If adult tickets cost \$10.25 each, the price of a adult tickets is $10.25a$. If child tickets cost \$7.75 each, the price of c child tickets is $7.75c$. The total price of the tickets is therefore $10.25a + 7.75c$. Since the cinema sold \$1470 worth of tickets, the other equation must be $10.25a + 7.75c = 1470$.

The correct choice is (**1**).

6. A function is linear if the graph of the function is a line. Here are the graphs for the four functions.

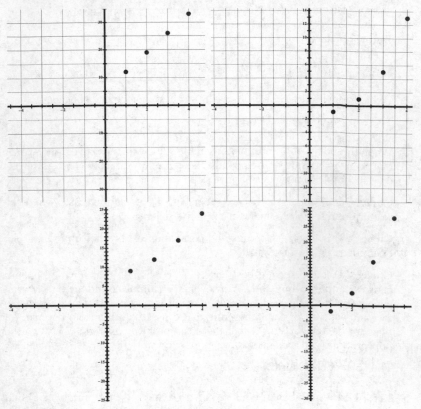

Only the graph of $f(x)$ is a line.

The correct choice is (**1**).

7. Test each of the four choices to see which, when put together with 7.2 and 7.6, produces an average between 7.0 and 7.8.

Testing choice (1):

$$\frac{7.2+7.6+6.2}{3} = \frac{21}{3} = 7$$

Testing choice (2):

$$\frac{7.2+7.6+7.3}{3} = \frac{22.1}{3} \approx 7.37$$

Testing choice (3):

$$\frac{7.2+7.6+8.6}{3} = \frac{23.4}{3} = 7.8$$

Testing choice (4):

$$\frac{7.2+7.6+8.8}{3} = \frac{23.6}{3} \approx 7.87$$

Only choice (2) is between 7.0 and 7.8. Choices (1) and (3) are incorrect because the inequality uses $<$ signs instead of $\le$ signs.

The correct choice is (**2**).

8. Since there are 60 seconds in a minute, you can convert seconds to minutes by dividing by 60. Therefore, 12.5 seconds is equal to $\frac{12.5}{60} \approx 0.2083$ minute. Since there are 3600 seconds in an hour, you can convert seconds to hours by dividing by 3600. Therefore, 12.5 seconds is equal to $\frac{12.5}{3600} \approx 0.0035$ hour.

The correct choice is (**1**).

9. The first step in solving this inequality is to distribute the 5 through the right-hand side.

$$3x + 2 \le 5x - 20$$

The next step is to subtract $5x$ from both sides of the inequality.

$$3x + 2 \le 5x - 20$$
$$-5x = -5x$$
$$-2x + 2 \le -20$$

The next step is to subtract 2 from both sides of the inequality.

$$-2x + 2 \le -20$$
$$-2 = -2$$
$$-2x \le -22$$

The last step is to divide both sides of the inequality by −2. When you divide an inequality by a negative number, the direction of the inequality sign must be reversed.

$$\frac{-2x}{-2} \geq \frac{-22}{-2}$$
$$x \geq 11$$

The correct choice is **(4)**.

10. The first step in simplifying this expression is to distribute the 3 through the first parentheses.

$$3x^2 - 3 - (x^2 - 7x + 10)$$

The next step is to distribute the negative (−) sign (which is really a −1) through the second parentheses.

$$3x^2 - 3 - x^2 + 7x - 10$$

Notice that there is a $+7x$ because $-1 \cdot (-7x) = +7x$.

The last step is to combine the like terms.

$$3x^2 - x^2 + 7x - 3 - 10$$
$$2x^2 + 7x - 13$$

The correct choice is **(2)**.

11. The range of a function is the set of all the possible y-coordinates of all the points on its graph. The graph of this function is a parabola with a minimum point at $(-1, -9)$. Since the parabola goes up forever, all y-coordinates greater than or equal to −9 will be represented on the graph.

For the TI-84:

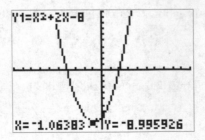

For the TI-Nspire:

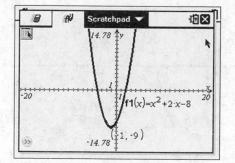

Another way to answer this question without using a graphing calculator is to recognize that the graph of the function will be a parabola that is opening up since the coefficient of the x^2-term is positive. This means that the vertex of the parabola will be the lowest point. So the y-coordinate of the vertex will be the lowest number in the range. The x-coordinate of the vertex can be determined with the formula:

$$x = \frac{-b}{2a} = \frac{-2}{2 \cdot 1} = \frac{-2}{2} = -1$$

The y-coordinate of the vertex can be determined by putting -1 into the function:

$$f(-1) = (-1)^2 + 2(-1) - 8 = 1 - 2 - 8 = -9$$

The correct choice is (2).

12. The zeros of a function are the x-values that make that function equal to zero. This question can be solved using the equation $x^2 - 5x - 6 = 0$.

$$x^2 - 5x - 6 = 0$$
$$(x - 6)(x + 1) = 0$$
$$x - 6 = 0 \quad \text{or} \quad x + 1 = 0$$
$$+6 = +6 \qquad\quad -1 = -1$$
$$x = 6 \quad \text{or} \qquad x = -1$$

This question can also be answered by graphing the function on the graphing calculator and finding the x coordinates of the x-intercepts.

For the TI-84:

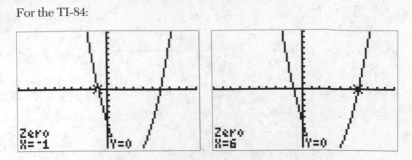

For the TI-Nspire:

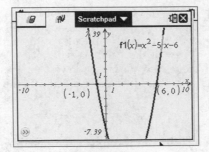

The correct choice is (1).

13. The common difference in an arithmetic sequence is the number that is added to each term to get the next term. If the first term is 4 and the common difference is 3, then the second term is 4 + 3 = 7, the third term is 7 + 3 = 10, the fourth term is 10 + 3 = 13, and the fifth term is 13 + 3 = 16.

The correct choice is (3).

14. One of the rules of exponents is that $x^{ab} = (x^a)^b$. By using this rule, the given function $C(t) = 10(1.029)^{24t}$ can be rewritten as $C(t) = 10(1.029^{24})^t$. Since $1.029^{24} \approx 1.986$, the function is equivalent to $C(t) = 10(1.986)^t$.

The correct choice is (3).

15. The total number of people in the 21 to 40 group is 30 + 12 + 8 = 50. Of those 50 people, 30 of them are for the candidate. To determine the percent for the candidate, calculate $\frac{30}{50} = 0.60$ equals 60%.

The correct choice is (4).

16. An expression is in vertex form if it looks like $a(x - h)^2 + k$. All four of the choices are in vertex form. Simplify each of the four choices to see which is equivalent to $x^2 - 12x + 7$.

    Choices (1) and (2):

    $(x - 6)^2 + 43 = (x - 6)(x - 6) + 43 = x^2 - 12x + 36 + 43 = x^2 - 12x + 79$

    Choices (3) and (4):

    $(x - 6)^2 - 29 = (x - 6)(x - 6) - 29 = x^2 - 12x + 36 - 29 = x^2 - 12x + 7$

    Choices (1) and (2) can be eliminated.

    When the equation of a parabola is in vertex form $a(x - h)^2 + k$, the coordinates of the vertex are (h, k). For the expression $(x - 6)^2 - 29$, the h is 6 and the k is -29. So the vertex is $(6, -29)$.

    The correct choice is **(3)**.

17. If \$500 is put into a savings account that earns 4% a year, it will earn $500(0.04) = 20$ dollars interest after the first year. So the total amount in the savings account will then be \$520. After the second year, it will earn $520(0.04) = 20.8$ dollars interest. So the total amount in the savings account will then be \$540.80. After the third year, it will earn $540.80(0.04) \approx 21.63$ dollars interest. So the total amount in the savings account will then be \$562.43.

    Answer choices (1), (3), and (4) all evaluate to \$562.43. However, answer choice (2) evaluates to \$442.37.

    The correct choice is **(2)**.

18. To find the equation of the line represented by the table, first choose any two rows from the table to represent two points on the line. For example, using the fourth and fifth rows means you are using the two points $(2, 4.6)$ and $(4, 5)$.

    Calculate the slope of the line.

    $$m = \frac{y_2 - y_1}{x_2 - x_1} = \frac{5 - 4.6}{4 - 2} = \frac{0.4}{2} = 0.2$$

The y-intercept of the line can be calculated with the formula $y = mx + b$, where m is the value that was just calculated and x and y are the x- and y-coordinates of either of the two points, for example (4, 5).

$$5 = 0.2 \cdot 4 + b$$
$$5 = 0.8 + b$$
$$-0.8 = -0.8$$
$$4.2 = b$$

The equation of the line represented by the table is $y = 0.2x + 4.2$.

The easiest way to find the solution to the system of equations $\begin{cases} 4y + 2x = 33.6 \\ y = 0.2x + 4.2 \end{cases}$ is to use the substitution method. Substitute the expression $0.2x + 4.2$ for the y in the first equation and solve for x.

$$4(0.2x + 4.2) + 2x = 33.6$$
$$0.8x + 16.8 + 2x = 33.6$$
$$2.8x + 16.8 = 33.6$$
$$-16.8 = -16.8$$
$$\frac{2.8x}{2.8} = \frac{16.8}{2.8}$$
$$x = 6$$

Substitute this value for x into either equation to solve for y.

$$y = 0.2(6) + 4.2 = 1.2 + 4.2 = 5.4$$

The correct choice is (**4**).

19.   There are several ways to solve this quadratic equation.

One way is to simplify the equation so there is a zero on the right-hand side. Then solve for x.

$$2(x + 2)^2 - 4 = 28$$
$$2(x^2 + 4x + 4) - 4 = 28$$
$$2x^2 + 8x + 8 - 4 = 28$$
$$2x^2 + 8x + 4 = 28$$
$$-28 = -28$$
$$2x^2 + 8x - 24 = 0$$
$$2(x^2 + 4x - 12) = 0$$
$$2(x + 6)(x - 2) = 0$$

$$x + 6 = 0 \quad \text{or} \quad x - 2 = 0$$
$$x = -6 \quad \text{or} \quad x = 2$$

A much shorter way to get to the same answer is to isolate the term in the parentheses first. Then solve for x.

$$2(x + 2)^2 - 4 = 28$$
$$+4 = +4$$
$$\frac{2(x+2)^2}{2} = \frac{32}{2}$$
$$(x + 2)^2 = 16$$
$$\sqrt{(x+2)^2} = \pm\sqrt{16}$$
$$x + 2 = \pm 4$$
$$-2 = -2$$
$$x = +4 - 2 = 2 \quad \text{or} \quad x = -4 - 2 = -6$$

A method that would not require algebra would be to test the numbers 2, 6, –2, and –6 in the equation since those are the only four numbers in any of the answer choices. If the number is part of the solution set, it will make the equation true.

Testing 2:

$$2(2 + 2)^2 - 4 \stackrel{?}{=} 28$$
$$2 \cdot 4^2 - 4 \stackrel{?}{=} 28$$
$$2 \cdot 16 - 4 \stackrel{?}{=} 28$$
$$32 - 4 \stackrel{?}{=} 28$$
$$28 \stackrel{\checkmark}{=} 28$$

Testing 6:

$$2(6 + 2)^2 - 4 \stackrel{?}{=} 28$$
$$2 \cdot 6^2 - 4 \stackrel{?}{=} 28$$
$$2 \cdot 36 - 4 \stackrel{?}{=} 28$$
$$72 - 4 \stackrel{?}{=} 28$$
$$68 \neq 28$$

Testing –2:

$$2(-2 + 2)^2 - 4 \stackrel{?}{=} 28$$
$$2 \cdot 0^2 - 4 \stackrel{?}{=} 28$$
$$2 \cdot 0 - 4 \stackrel{?}{=} 28$$
$$0 - 4 \stackrel{?}{=} 28$$
$$-4 \neq 28$$

Testing –6:

$$2(-6 + 2)^2 - 4 \overset{?}{=} 28$$
$$2 \cdot (-4)^2 - 4 \overset{?}{=} 28$$
$$2 \cdot 16 - 4 \overset{?}{=} 28$$
$$32 - 4 \overset{?}{=} 28$$
$$28 \overset{\checkmark}{=} 28$$

The correct choice is **(3)**.

20. Based on the dot plot, the numbers in the set are

$$\{0, 0, 1, 1, 2, 2, 2, 2, 2, 3, 3, 3, 3, 3, 4, 4, 4, 5, 5, 6\}$$

There are 20 numbers altogether.

The median is the middle number when the numbers are arranged from lowest to highest. With 20 numbers, there are two numbers in the middle. So the median is the average of those two numbers. Since the two middle numbers are the 10th and the 11th numbers and those numbers are both 3, the median is 3. So choice (1) can be eliminated.

To get the interquartile range, split the data into two halves and find the median of each half. The half with the lower 10 numbers is $\{0, 0, 1, 1, 2, 2, 2, 2, 2, 3\}$. The median of these 10 numbers is the average of the 5th and 6th numbers, which are both 2. So the value of the first quartile is 2. The half with the higher 10 numbers is $\{3, 3, 3, 3, 4, 4, 4, 5, 5, 6\}$. The median of these 10 numbers is the average of the 5th and 6th numbers, which are both 4. So the value of the third quartile is 4. The difference between the values of the third quartile and the first quartile is called the interquartile range. For this data set, the interquartile range is $4 - 2 = 2$. So choice (2) can be eliminated.

The mean is the average of the 20 numbers. For this set, the sum of the numbers is 55. So the mean is $\frac{55}{20} = 2.75$ which is not equal to 3.

This data can also be put into the graphing calculator to find the mean, median, first quartile, and third quartile.

For the TI-84:

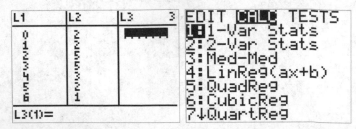

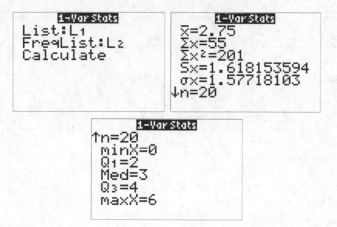

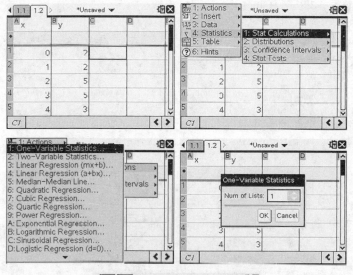

For the TI-Nspire:

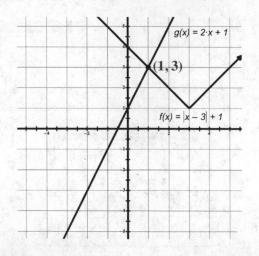

The correct choice is **(3)**.

21. Test the four answer choices. Start with the largest value, choice (4), to see which is the largest among the choices that makes $f(x) > g(x)$.

Testing choice (4):

$f(10) = 5 \cdot 10^4 + 30 \cdot 10^2 + 9 = 53{,}009$
$g(10) = 3^{10} = 59{,}049$
$f(10) < g(10)$

Testing choice (3):

$f(10) = 5 \cdot 9^4 + 30 \cdot 9^2 + 9 = 35{,}244$
$g(10) = 3^9 = 19{,}683$
$f(9) > g(9)$

The correct choice is **(3)**.

22. On the same set of axes, graph both functions.

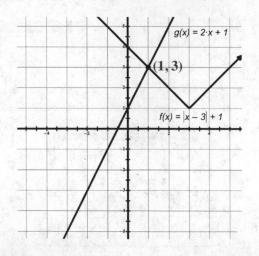

The graphs intersect at the point $(1, 3)$. This means that $f(1) = g(1) = 3$. So the solution to $f(x) = g(x)$ is 1.

The graphs and intersection can also be found on the graphing calculator.

For the TI-84:

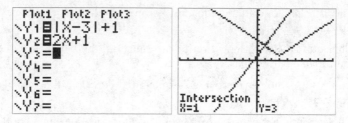

For the TI-Nspire:

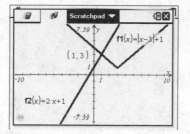

The correct choice is (2).

23. It is possible to have exactly 3 ounces of frozen yogurt, but it is also possible to have 3.1 ounces of frozen yogurt. It is not possible to have a negative amount of frozen yogurt.

Choice (1) says that the amount of frozen yogurt must be an exact integer, positive or negative, like 1, 2, 3, –1, –2, or –3 ounces.

Choice (2) says the amount of frozen yogurt can be a fraction or a decimal, including negative fractions or decimals.

Choice (3) says the amount of frozen yogurt cannot be negative but must be an integer like 1, 2, or 3 ounces and cannot be something in between two exact ounces.

The correct choice is (4).

24. To set up this equation, call the first consecutive integer x and the next consecutive integer $x + 1$. The equation becomes:

$$x(x + 1) = 156$$
$$x^2 + x = 156$$
$$x^2 + x - 156 = 0$$

This is called a quadratic equation because the highest exponent on the variable x is 2.

Choice (1) is not correct. In a linear equation, the highest exponent on the variable is 1, such as $2x + 1 = 11$.

Choice (3) is not correct. In an exponential equation, the variable is an exponent, such as $3^x + 1 = 82$.

Choice (4) is not correct. An absolute value equation must contain the absolute value sign, such as $|x + 2| - 3 = 7$.

The correct choice is **(2)**.

PART II

25. Replace the expression for $f(x)$ into the $g(x)$ equation and simplify.

$$g(x) = 2[f(x)]^2 - 1$$
$$g(x) = 2(2x + 1)^2 - 1$$
$$g(x) = 2(2x + 1)(2x + 1) - 1$$
$$g(x) = 2(4x^2 + 4x + 1) - 1$$
$$g(x) = 8x^2 + 8x + 2 - 1$$
$$g(x) = 8x^2 + 8x + 1$$

26. $3\sqrt{2} \cdot 8\sqrt{18} = 24\sqrt{36} = 24 \cdot 6 = 144$. Since 144 is a rational number, the product is rational.

27. One way to draw the graph is to make a table of values.

x	y
–3	$(-3)^2 - 4(-3) - 1 = 20$
–2	$(-2)^2 - 4(-2) - 1 = 11$
–1	$(-1)^2 - 4(-1) - 1 = 4$
0	$0^2 - 4(0) - 1 = -1$
1	$1^2 - 4(1) - 1 = -4$
2	$2^2 - 4(2) - 1 = -5$
3	$3^2 - 4(3) - 1 = -4$

The graph looks like this.

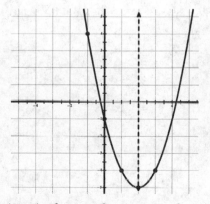

The vertex is at $(2, -5)$. The axis of symmetry is the vertical line through the vertex and has the equation $x = 2$.

28. Amy is correct. One way to verify that she is correct is to substitute the values $\frac{7}{2}$ and -6 into the equation to see if they make it true.

Testing $\frac{7}{2}$:

$$2\left(\frac{7}{2}\right)^2 + 5\left(\frac{7}{2}\right) - 42 \stackrel{?}{=} 0$$

$$2\left(\frac{49}{4}\right) + 5\left(\frac{7}{2}\right) - 42 \stackrel{?}{=} 0$$

$$\frac{98}{4} + \frac{35}{2} - 42 \stackrel{?}{=} 0$$

$$\frac{98}{4} + \frac{70}{4} - 42 \stackrel{?}{=} 0$$

$$\frac{168}{4} - 42 \stackrel{?}{=} 0$$

$$42 - 42 \stackrel{?}{=} 0$$

$$0 \stackrel{\checkmark}{=} 0$$

Testing -6:

$$2(-6)^2 + 5(-6) - 42 \stackrel{?}{=} 0$$
$$2(36) - 30 - 42 \stackrel{?}{=} 0$$
$$72 - 30 - 42 \stackrel{?}{=} 0$$
$$0 \stackrel{\checkmark}{=} 0$$

Another way to do this would be to solve the equation by factoring.

$$2x^2 + 5x - 42 = 0$$
$$(2x - 7)(x + 6) = 0$$
$$2x - 7 = 0 \quad \text{or} \quad x + 6 = 0$$
$$2x = 7 \quad \text{or} \quad x = -6$$
$$x = \frac{7}{2} \quad \text{or} \quad x = -6$$

A third way to do this would be to use the graphing calculator to find the x-intercepts of the graph of the equation $y = 2x^2 + 5x - 42$.

For the TI-84:

For the TI-Nspire:

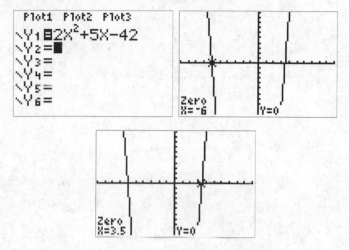

29. Determine the slope of the line.

$$m = \frac{y_2 - y_1}{x_2 - x_1} = \frac{1 - 4}{6 - (-3)} = \frac{-3}{9} = -\frac{1}{3}$$

Use the point-slope formula $y - y_1 = m(x - x_1)$ with $m = -\frac{1}{3}$ and $(x_1, y_1) = (-3, 4)$. $y - 4 = -\frac{1}{3}(x - (-3)) = y - 4 = -\frac{1}{3}(x + 3)$. This is Sue's answer.

Simplify Sue's equation by adding 4 to both sides. Then simplify the right-hand side to get Kathy's answer.

$$y - 4 = -\frac{1}{3}(x + 3)$$
$$+4 = +4$$
$$y = -\frac{1}{3}(x + 3) + 4$$
$$y = -\frac{1}{3}x + \left(-\frac{1}{3}\right)3 + 4$$
$$y = -\frac{1}{3}x - 1 + 4$$
$$y = -\frac{1}{3}x + 3$$

To do this without using the point-slope formula, instead use the formula $y = mx + b$, where $m = -\frac{1}{3}$ and x and y are the x- and y-coordinates of either of the given points, like $(6,1)$. Solve for b.

$$y = mx + b$$
$$1 = -\frac{1}{3} \cdot 6 + b$$
$$1 = -2 + b$$
$$+2 = +2$$
$$3 = b$$

So the equation is what Kathy got, $y = -\frac{1}{3}x + 3$. Simplifying Sue's response leads to the same equation as Kathy found.

30. The slope of a line that passes through $(3, 4)$ and $(7, 6)$ is

$$m = \frac{y_2 - y_1}{x_2 - x_1} = \frac{6 - 4}{7 - 3} = \frac{2}{4} = \frac{1}{2}$$

If 2 inches of snow fall in 4 hours, $\frac{1}{2}$ is the number of inches of snow that falls in one hour.

31. There are two ways to isolate the n variable.

The first method uses the distributive property.

$$S = 180(n - 2)$$
$$S = 180n - 360$$
$$+360 = +360$$
$$S + 360 = 180n$$
$$\frac{S + 360}{180} = \frac{180n}{180}$$
$$\frac{S + 360}{180} = n$$

The second method uses the division property of equality.

$$S = 180(n - 2)$$
$$\frac{S}{180} = \frac{180(n - 2)}{180}$$
$$\frac{S}{180} = n - 2$$
$$+2 = +2$$
$$\frac{S}{180} + 2 = n$$

32. The equation for $g(x)$ is $g(x) = f(x) - 4$. The graph of $g(x)$ is like the graph of $f(x)$ but translated down 4 units. In general, a translation up by a units of the graph of $f(x)$ can be represented as $f(x) + a$ and a translation down by a units of the graph of $f(x)$ can be represented as $f(x) - a$.

PART III

33. After 1 second, the height of the object is

$$H(1) = -16 \cdot 1^2 + 144 = -16 + 144 = 128$$

After 2 seconds, the height of the object is

$$H(2) = -16 \cdot 2^2 + 144 = -16 \cdot 4 + 144 = 80$$

The number of feet that the object dropped between 1 second and 2 seconds is $128 - 80 = 48$ feet.

The object reaches the ground when its height off the ground is 0. To find out when this is, set $0 = H(t)$ or $0 = -16t^2 + 144$. Solve this quadratic equation.

$$0 = -16t^2 + 144$$
$$-144 = -144$$
$$-144 = -16t^2$$

$$\frac{-144}{-16} = \frac{-16t^2}{-16}$$

$$9 = t^2$$
$$\pm 3 = t$$

Since this is an amount of time, the negative answer should be rejected. The only solution is $t = 3$ seconds for the object to reach the ground.

34. The inequalities are $x + y > 8$ and $2x + y < 14$. To graph the first inequality, isolate the y to make it $y > -x + 8$. Because it is a $>$ sign instead of a $\geq$ sign, make the line $y = -x + 8$ a dotted line with a y-intercept of $(0, 8)$ and a slope of -1.

To decide which side of the dotted line to shade, test if the ordered pair $(0, 0)$ makes the inequality true or false.

$$0 \overset{?}{>} -0 + 8$$
$$0 > 8$$

Since this is not true, shade the side that does *not* contain the point $(0, 0)$.

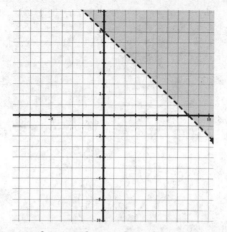

To graph the second inequality, isolate the y to make it $y < -2x + 14$. Because it is a $<$ sign instead of a $\leq$ sign, make the line $y = -2x + 14$ a dotted line with a y-intercept of $(0, 14)$ and a slope of -2.

To decide which side of the dotted line to shade, test if the ordered pair $(0, 0)$ makes the inequality true or false.

$$0 \overset{?}{<} -2 \cdot 0 + 14$$
$$0 \overset{?}{<} 0 + 14$$
$$0 > 14$$

Since this is true, shade the side that contains the point $(0, 0)$.

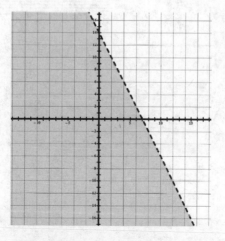

When both inequalities are shown on the same coordinate axes, the graph looks like this.

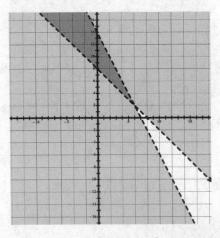

The inequalities can also be graphed with the TI graphing calculators.

For the TI-84:

For the TI-Nspire:

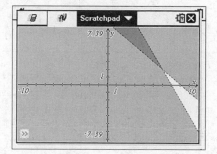

Since the point (6, 2) is on the intersection of the borderlines and not in the double-shaded region, Kai is not correct. Another way to check this is to see if 6 + 2 > 8. This inequality is not true since 6 + 2 = 8.

35. After 92 minutes, the plane has spent 32 minutes reaching cruising altitude and 92 − 32 = 60 minutes at cruising altitude. The plane has traveled 192 miles before reaching cruising altitude and 762 − 192 = 570 miles at cruising altitude. The speed of the plane at cruising altitude is therefore $\frac{570}{60} = 9.5$ miles per minute.

When the plane has flown for x minutes at cruising altitude, it has traveled 9.5x miles. So the equation is $y = 9.5x$.

Two hours is 120 minutes. During the first 32 minutes, the plane traveled 192 miles before reaching cruising altitude. The amount of time the plane spent at cruising altitude is 120 − 32 = 88 minutes. The distance traveled at cruising altitude is 9.5 · 88 = 836 miles at cruising altitude. Add the 192 miles already traveled. After 2 hours, the airplane has traveled 192 + 835 = 1,028 miles.

36. The graph of $g(x)$ is a line with y-intercept (0, 1) and slope $\frac{1}{2}$. Since $f(x)$ is a piecewise function, the graph is a line to the left of and at −1 and is a parabola to the right of −1. Making a chart helps produce the graph. In a piecewise function, it is useful to determine the function value for both pieces at the boundary, which is −1 in this case.

x	$g(x)$	$f(x)$
-4	$\frac{1}{2}(-4) + 1 = -1$	$2(-4) + 1 = -7$
-3	$\frac{1}{2}(-3) + 1 = -0.5$	$2(-3) + 1 = -5$
-2	$\frac{1}{2}(-2) + 1 = 0$	$2(-2) + 1 = -3$
-1	$\frac{1}{2}(-1) + 1 = 0.5$	$2(-1) + 1 = -1$ Graphed with closed circle
-1		$2 - (-1)^2 = -1$ Graphed with open circle
0	$\frac{1}{2}(0) + 1 = 1$	$2 - 0^2 = 2$
1	$\frac{1}{2}(-1) + 1 = 1.5$	$2 - 1^2 = 1$
2	$\frac{1}{2}(2) + 1 = 2$	$2 - 2^2 = -2$
3	$\frac{1}{2}(3) + 1 = 2.5$	$2 - 3^2 = -7$
4	$\frac{1}{2}(4) + 1 = 3$	$2 - 4^2 = -14$

The graph looks like this.

For the TI-84:

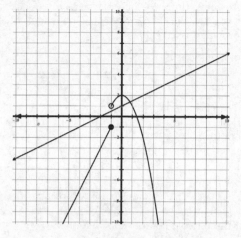

For the TI-Nspire:

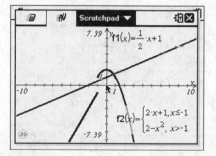

Since the graph of $g(x)$ seems to intersect the parabola piece of $f(x)$ in one place, there is just one value where $f(x) = g(x)$.

PART IV

37. If x is the price of one package of cupcakes and y is the price of one package of brownies, the price of 3 packages of cupcakes and 2 packages of brownies is $3x + 2y$. The price of 2 packages of cupcakes and 4 packages of brownies is $2x + 4y$.

The system of equations is

$$\begin{cases} 3x + 2y = 19 \\ 2x + 4y = 24 \end{cases}$$

These equations can be graphed by first converting each of them into slope-intercept form.

$$\begin{cases} y = -\dfrac{3}{2}x + \dfrac{19}{2} \\ y = -\dfrac{1}{2}x + 6 \end{cases}$$

The graph of these two equations look like this.

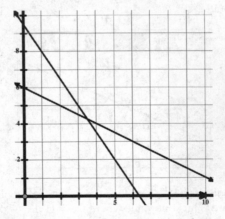

The x-coordinate of the point of intersection is the price of a package of cupcakes. The y-coordinate of the point of intersection is the price of a package of brownies.

Using a graphing calculator, the intersection point can be found to be $(3.5, 4.25)$. So a package of cupcakes costs \$3.50 and a package of brownies costs \$4.25.

For the TI-84:

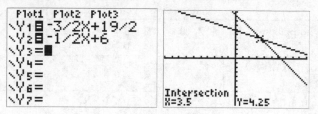

For the TI-Nspire:

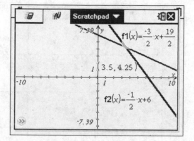

The system can be solved without graphing by using algebra.

$$3x + 2y = 19$$
$$2x + 4y = 24$$

Multiply the top equation by –2 so that the y-variables in the two equations will be the same number but with opposite signs.

$$-6x - 4y = -38$$
$$2x + 4y = 24$$

Add the two equations, and solve the resulting equation for x

$$-4x = -14$$

$$\frac{-4x}{-4} = \frac{-14}{-4}$$

$$x = 3.5$$

Substitute $x = 3.5$ into either of the original equations, and solve for y.

$$2(3.5) + 4y = 24$$
$$7 + 4y = 24$$
$$-7 = -7$$
$$4y = 17$$

$$\frac{4y}{4} = \frac{17}{4}$$

$$y = 4.25$$

Topic	Question Numbers	Number of Points	Your Points	Your Percentage
1. Polynomials	1, 2, 10	2 + 2 + 2 = 6		
2. Properties of Algebra	31	2		
3. Functions	3, 6, 11, 12, 23, 25, 32	2 + 2 + 2 + 2 + 2 + 2 + 4 = 16		
4. Creating and Interpreting Equations	24, 29, 30, 35	2 + 2 + 2 + 4 = 10		
5. Inequalities	7, 9, 21	2 + 2 + 2 = 6		
6. Sequences and Series	13	2		
7. Systems of Equations	5, 18, 22, 34, 37	2 + 2 + 2 + 4 + 6 = 16		
8. Quadratic Equations and Factoring	16, 19, 28, 33	2 + 2 + 2 + 4 = 10		
9. Regression	4	2		
10. Exponential Equations	14, 17	2 + 2 = 4		
11. Graphing	27, 36	2 + 4 = 6		
12. Statistics	15, 20	2 + 2 = 4		
13. Number Properties	8, 26	2 + 2 = 4		

HOW TO CONVERT YOUR RAW SCORE TO YOUR ALGEBRA I REGENTS EXAMINATION SCORE

The accompanying conversion chart must be used to determine your final score on the June 2016 Regents Examination in Algebra I. To find your final exam score, locate in the column labeled "Raw Score" the total number of points you scored out of a possible 86 points. Since partial credit is allowed in Parts II, III, and IV of the test, you may need to approximate the credit you would receive for a solution that is not completely correct. Then locate in the adjacent column to the right the scale score that corresponds to your raw score. The scale score is your final Algebra I Regents Examination score.

Regents Examination in Algebra I—June 2016
Chart for Converting Total Test Raw Scores to Final
Examination Scores (Scaled Scores)

Raw Score	Scale Score	Performance Level	Raw Score	Scale Score	Performance Level	Raw Score	Scale Score	Performance Level
86	100	5	57	81	4	28	66	3
85	99	5	56	81	4	27	65	3
84	98	5	55	81	4	26	64	2
83	96	5	54	81	4	25	63	2
82	95	5	53	80	4	24	61	2
81	94	5	52	80	4	23	60	2
80	93	5	51	80	4	22	58	2
79	92	5	50	79	3	21	57	2
78	92	5	49	79	3	20	55	2
77	91	5	48	79	3	19	53	1
76	90	5	47	78	3	18	51	1
75	89	5	46	78	3	17	49	1
74	89	5	45	78	3	16	47	1
73	88	5	44	77	3	15	45	1
72	87	5	43	77	3	14	43	1
71	86	5	42	77	3	13	41	1
70	86	5	41	76	3	12	38	1
69	86	5	40	76	3	11	36	1
68	86	5	39	75	3	10	33	1
67	85	5	38	74	3	9	30	1
66	84	4	37	74	3	8	28	1
65	84	4	36	73	3	7	25	1
64	84	4	35	72	3	6	21	1
63	83	4	34	72	3	5	18	1
62	83	4	33	71	3	4	15	1
61	82	4	32	70	3	3	11	1
60	82	4	31	69	3	2	8	1
59	82	4	30	68	3	1	4	1
58	82	4	29	67	3	0	0	1

Examination
August 2016
Algebra I

HIGH SCHOOL MATH REFERENCE SHEET

Conversions

1 inch = 2.54 centimeters

1 meter = 39.37 inches

1 mile = 5280 feet

1 mile = 1760 yards

1 mile = 1.609 kilometers

1 cup = 8 fluid ounces

1 pint = 2 cups

1 quart = 2 pints

1 gallon = 4 quarts

1 gallon = 3.785 liters

1 liter = 0.264 gallon

1 kilometer = 0.62 mile

1 pound = 16 ounces

1 pound = 0.454 kilogram

1 kilogram = 2.2 pounds

1 ton = 2000 pounds

1 liter = 1000 cubic centimeters

Formulas

Triangle	$A = \dfrac{1}{2}bh$
Parallelogram	$A = bh$
Circle	$A = \pi r^2$
Circle	$C = \pi d \text{ or } C = 2\pi r$

Formulas (continued)

General Prisms	$V = Bh$
Cylinder	$V = \pi r^2 h$
Sphere	$V = \frac{4}{3}\pi r^3$
Cone	$V = \frac{1}{3}\pi r^2 h$
Pyramid	$V = \frac{1}{3}Bh$
Pythagorean Theorem	$a^2 + b^2 = c^2$
Quadratic Formula	$x = \dfrac{-b \pm \sqrt{b^2 - 4ac}}{2a}$
Arithmetic Sequence	$a_n = a_1 + (n-1)d$
Geometric Sequence	$a_n = a_1 r^{n-1}$
Geometric Series	$S_n = \dfrac{a_1 - a_1 r^n}{1 - r}$ where $r \neq 1$
Radians	1 radian $= \dfrac{180}{\pi}$ degrees
Degrees	1 degree $= \dfrac{\pi}{180}$ radians
Exponential Growth/Decay	$A = A_0 e^{k(t - t_0)} + B_0$

PART I

Answer all **24 questions** in this part. Each correct answer will receive 2 credits. No partial credit will be allowed. For each statement or question, write in the space provided the numeral preceding the word or expression that best completes the statement or answers the question. [48 credits]

1 The graph below shows the distance in miles, m, hiked from a camp in h hours.

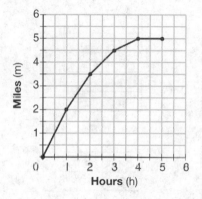

Which hourly interval had the greatest rate of change?

(1) hour 0 to hour 1   (3) hour 2 to hour 3

(2) hour 1 to hour 2   (4) hour 3 to hour 4   1 _____

2 The solution of an equation with two variables, x and y, is

(1) the set of all x values that make $y = 0$

(2) the set of all y values that make $x = 0$

(3) the set of all ordered pairs, (x, y), that make the equation true

(4) the set of all ordered pairs, (x, y), where the graph of the equation crosses the y-axis   2 _____

3 Which statistic can *not* be determined from a box plot representing the scores on a math test in Mrs. DeRidder's algebra class?

(1) the lowest score
(2) the median score
(3) the highest score
(4) the score that occurs most frequently

3 _____

4 Which chart could represent the function $f(x) = -2x + 6$?

x	f(x)
0	6
2	10
4	14
6	18

(1)

x	f(x)
0	8
2	10
4	12
6	14

(3)

x	f(x)
0	4
2	6
4	8
6	10

(2)

x	f(x)
0	6
2	2
4	-2
6	-6

(4)

4 _____

5 If $f(n) = (n - 1)^2 + 3n$, which statement is true?

(1) $f(3) = -2$ (3) $f(-2) = -15$
(2) $f(-2) = 3$ (4) $f(-15) = -2$

5 _____

6 The table below shows 6 students' overall averages and their averages in their math class.

Overall Student Average	92	98	84	80	75	82
Math Class Average	91	95	85	85	75	78

If a linear model is applied to these data, which statement best describes the correlation coefficient?

(1) It is close to –1.   (3) It is close to 0.

(2) It is close to 1.   (4) It is close to 0.5.   6 _____

7 What is the solution to $2h + 8 > 3h - 6$?

(1) $h < 14$ (3) $h > 14$

(2) $h < \dfrac{14}{5}$ (4) $h > \dfrac{14}{5}$ 7 _____

8 Which expression is equivalent to $36x^2 - 100$?

(1) $4(3x - 5)(3x - 5)$ (3) $2(9x - 25)(9x - 25)$

(2) $4(3x + 5)(3x - 5)$ (4) $2(9x + 25)(9x - 25)$ 8 _____

9 Patricia is trying to compare the average rainfall of New York to that of Arizona. A comparison between these two states for the months of July through September would be best measured in

(1) feet per hour   (3) inches per month

(2) inches per hour   (4) feet per month   9 _____

10 Which function defines the sequence –6, –10, –14, –18, ..., where $f(6) = -26$?
(1) $f(x) = -4x - 2$
(2) $f(x) = 4x - 2$
(3) $f(x) = -x + 32$
(4) $f(x) = x - 26$

10 _____

11 Which function has the greatest y-intercept?

(1) $f(x) = 3x$
(2) $2x + 3y = 12$
(3) the line that has a slope of 2 and passes through $(1, -4)$

(4)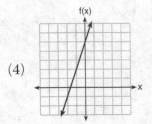

11 _____

12 What is the product of $2x + 3$ and $4x^2 - 5x + 6$?

(1) $8x^3 - 2x^2 + 3x + 18$
(2) $8x^3 - 2x^2 - 3x + 18$
(3) $8x^3 + 2x^2 - 3x + 18$
(4) $8x^3 + 2x^2 + 3x + 18$

12 _____

13 The height of a rocket, at selected times, is shown in the table below.

Time (sec)	0	1	2	3	4	5	6	7
Height (ft)	180	260	308	324	308	260	180	68

Based on these data, which statement is *not* a valid conclusion?

(1) The rocket was launched from a height of 180 feet.
(2) The maximum height of the rocket occurred 3 seconds after launch.
(3) The rocket was in the air approximately 6 seconds before hitting the ground.
(4) The rocket was above 300 feet for approximately 2 seconds.

13 _____

14 A parking garage charges a base rate of $3.50 for up to 2 hours, and an hourly rate for each additional hour. The sign below gives the prices for up to 5 hours of parking.

Parking Rates
2 hours $3.50
3 hours $9.00
4 hours $14.50
5 hours $20.00

Which linear equation can be used to find x, the additional hourly parking rate?

(1) $9.00 + 3x = 20.00$ (3) $2x + 3.50 = 14.50$
(2) $9.00 + 3.50x = 20.00$ (4) $2x + 9.00 = 14.50$

14 _____

15 Which function has a constant rate of change equal to −3?

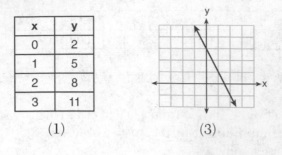

x	y
0	2
1	5
2	8
3	11

(1)

(3)

{(1, 5), (2, 2), (3, −5), (4, 4)} $2y = -6x + 10$

(2)   (4)

15 _____

16 Kendal bought x boxes of cookies to bring to a party. Each box contains 12 cookies. She decides to keep two boxes for herself. She brings 60 cookies to the party. Which equation can be used to find the number of boxes, x, Kendal bought?

(1) $2x - 12 = 60$ (3) $12x - 24 = 60$
(2) $12x - 2 = 60$ (4) $24 - 12x = 60$

16 _____

17 The table below shows the temperature, $T(m)$, of a cup of hot chocolate that is allowed to chill over several minutes, m.

Time, m (minutes)	0	2	4	6	8
Temperature, T(m) (°F)	150	108	78	56	41

Which expression best fits the data for $T(m)$?

(1) $150(0.85)^m$ (3) $150(0.85)^{m-1}$
(2) $150(1.15)^m$ (4) $150(1.15)^{m-1}$

17 _____

18 As x increases beyond 25, which function will have the largest value?

(1) $f(x) = 1.5^x$ (3) $h(x) = 1.5x^2$

(2) $g(x) = 1.5x + 3$ (4) $k(x) = 1.5x^3 + 1.5x^2$ 18 _____

19 What are the solutions to the equation $3x^2 + 10x = 8$?

(1) $\dfrac{2}{3}$ and -4 (3) $\dfrac{4}{3}$ and -2

(2) $-\dfrac{2}{3}$ and 4 (4) $-\dfrac{4}{3}$ and 2 19 _____

20 An online company lets you download songs for $0.99 each after you have paid a $5 membership fee. Which domain would be most appropriate to calculate the cost to download songs?

(1) rational numbers greater than zero
(2) whole numbers greater than or equal to one
(3) integers less than or equal to zero
(4) whole numbers less than or equal to one 20 _____

21 The function $f(x) = 3x^2 + 12x + 11$ can be written in vertex form as

(1) $f(x) = (3x + 6)^2 - 25$
(2) $f(x) = 3(x + 6)^2 - 25$
(3) $f(x) = 3(x + 2)^2 - 1$
(4) $f(x) = 3(x + 2)^2 + 7$ 21 _____

22 A system of equations is given below.

$$x + 2y = 5$$
$$2x + y = 4$$

Which system of equations does *not* have the same solution?

(1) $3x + 6y = 15$ (3) $x + 2y = 5$
     $2x + y = 4$ $6x + 3y = 12$

(2) $4x + 8y = 20$ (4) $x + 2y = 5$
     $2x + y = 4$ $4x + 2y = 12$ 22 _____

23 Based on the graph below, which expression is a possible factorization of $p(x)$?

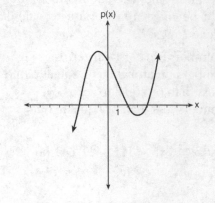

(1) $(x + 3)(x - 2)(x - 4)$
(2) $(x - 3)(x + 2)(x + 4)$
(3) $(x + 3)(x - 5)(x - 2)(x - 4)$
(4) $(x - 3)(x + 5)(x + 2)(x + 4)$ 23 _____

24 Milton has his money invested in a stock portfolio. The value, $v(x)$, of his portfolio can be modeled with the function $v(x) = 30,000(0.78)^x$, where x is the number of years since he made his investment. Which statement describes the rate of change of the value of his portfolio?

(1) It decreases 78% per year.
(2) It decreases 22% per year.
(3) It increases 78% per year.
(4) It increases 22% per year.   24 _____

PART II

Answer all 8 questions in this part. Each correct answer will receive 2 credits. Clearly indicate the necessary steps, including appropriate formula substitutions, diagrams, graphs, charts, etc. For all questions in this part, a correct numerical answer with no work shown will receive only 1 credit.   [16 credits]

25 Graph the function $y = -\sqrt{x+3}$ on the set of axes below.

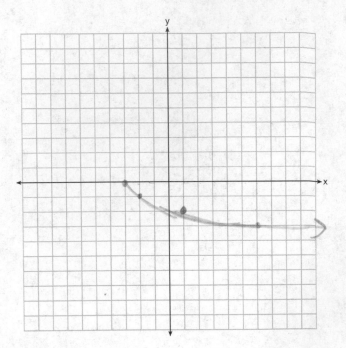

26 Richard is asked to transform the graph of $b(x)$ below.

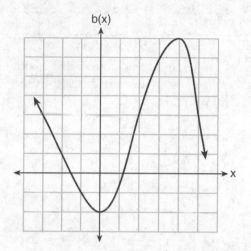

The graph of $b(x)$ is transformed using the equation
$h(x) = b(x - 2) - 3$. Describe how the graph of $b(x)$
changed to form the graph of $h(x)$.

right 2, down 3

27 Consider the pattern of squares shown below:

2   4   8

Which type of model, linear or exponential, should be used to determine how many squares are in the nth pattern? Explain your answer.

exponential, each time the squares are being multiplied by 2.

28 When multiplying polynomials for a math assignment, Pat found the product to be $-4x + 8x^2 - 2x^3 + 5$. He then had to state the leading coefficient of this polynomial. Pat wrote down -4. Do you agree with Pat's answer? Explain your reasoning.

No, because it is not in standard form. If it was the leading coefficient would be 2.

29 Is the sum of $3\sqrt{2}$ and $4\sqrt{2}$ rational or irrational? Explain your answer.

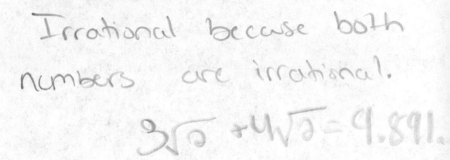

Irrational because both numbers are irrational.

$3\sqrt{2} + 4\sqrt{2} = 9.891.$

30 The graph below shows two functions, $f(x)$ and $g(x)$. State all the values of x for which $f(x) = g(x)$.

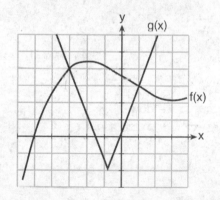

$X = 1 \quad X = -3$

31 Find the zeros of $f(x) = (x - 3)^2 - 49$, algebraically.

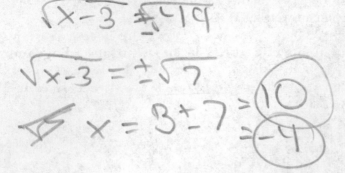

$\sqrt{x-3} = \pm\sqrt{49}$

$\sqrt{x-3} = \pm\sqrt{7}$

$x = 3 \pm 7 = \boxed{10}$
$= \boxed{-4}$

32 Solve the equation below for x in terms of a.

$$4(ax + 3) - 3ax = 25 + 3a$$

$4ax + 12 - 3ax = 25 + 3a$

$ax + 12 = 25 + 3a$

$\dfrac{ax + 13}{a} = \dfrac{3a}{a}$

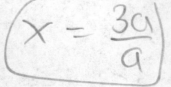

$\boxed{x = \dfrac{3a}{a}}$

PART III

Answer all **4** questions in this part. Each correct answer will receive **4** credits. Clearly indicate the necessary steps, including appropriate formula substitutions, diagrams, graphs, charts, etc. For all questions in this part, a correct numerical answer with no work shown will receive only **1** credit. [16 credits]

33 The data table below shows the median diameter of grains of sand and the slope of the beach for 9 naturally occurring ocean beaches.

Median Diameter of Grains of Sand, in Millimeters (x)	0.17	0.19	0.22	0.235	0.235	0.3	0.35	0.42	0.85
Slope of Beach, in Degrees (y)	0.63	0.7	0.82	0.88	1.15	1.5	4.4	7.3	11.3

Write the linear regression equation for this set of data, rounding all values to the *nearest thousandth*.

$$y = 17.159x - 2.476$$

Using this equation, predict the slope of a beach, to the *nearest tenth of a degree*, on a beach with grains of sand having a median diameter of 0.65 mm.

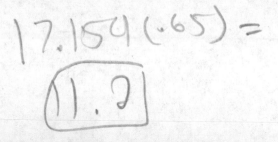

$$17.159(.65) =$$

$$\boxed{11.2}$$

34 Shawn incorrectly graphed the inequality $-x - 2y \geq 8$ as shown below.

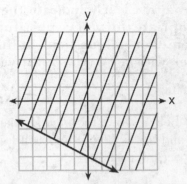

Explain Shawn's mistake.

He didn't flip the sign.

Graph the inequality correctly on the set of axes below.

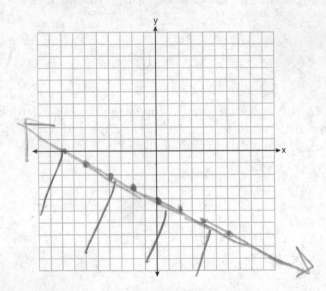

35 A drama club is selling tickets to the spring musical.
The auditorium holds 200 people. Tickets cost $12
at the door and $8.50 if purchased in advance. The
drama club has a goal of selling at least $1000 worth
of tickets to Saturday's show.

· Write a system of inequalities that can be used to
model this scenario.

$$12x + 8.50y \geq 1000$$

$$x + y \leq 200$$

If 50 tickets are sold in advance, what is the minimum
number of tickets that must be sold at the door so that
the club meets its goal? Justify your answer.

$$12x + 8.50(50) \geq 1000$$

$$12x + 425 \geq 1000$$

$$ -425 \quad\quad -425$$

$$\frac{12x}{12} \geq \frac{575}{12}$$

$$\boxed{x \geq 48}$$

36 Janice is asked to solve $0 = 64x^2 + 16x - 3$. She begins
    the problem by writing the following steps:

    Line 1 $0 = 64x^2 + 16x - 3$
    Line 2 $0 = B^2 + 2B - 3$
    Line 3 $0 = (B + 3)(B - 1)$

Use Janice's procedure to solve the equation for x.

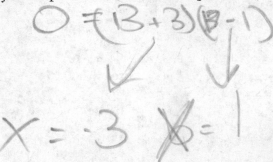

$$0 = (B + 3)(B - 1)$$

$$x = -3 \quad \cancel{B} = 1$$

Explain the method Janice used to solve the quadratic
equation.

Janice used the
method of factoring.

PART IV

**Answer the question in this part. A correct answer will
receive 6 credits. Clearly indicate the necessary steps, including
appropriate formula substitutions, diagrams, graphs,
charts, etc. A correct numerical answer with no work shown
will receive only 1 credit.** [6 credits]

37 For a class picnic, two teachers went to the same store
to purchase drinks. One teacher purchased 18 juice
boxes and 32 bottles of water, and spent $19.92. The
other teacher purchased 14 juice boxes and 26 bottles
of water, and spent $15.76.

Write a system of equations to represent the costs of a
juice box, j, and a bottle of water, w.

$$18j + 32w = 19.92$$
$$14j + 26w = 15.76$$

Kara said that the juice boxes might have cost 52 cents
each and that the bottles of water might have cost 33
cents each. Use your system of equations to justify that
Kara's prices are *not* possible.

$$18(52) + 32(33) = 19.92$$
$$14(52) + 26(33) = 15.83$$

Question 37 is continued on the next page.

Question 37 continued

Solve your system of equations to determine the actual cost, in dollars, of each juice box and each bottle of water.

$$14(18j + 32w) = (19.92)$$

$$-18(14j + 26w) = (15.76)$$

$$252j + 448w = 278.88$$

$$-252j - 468w = 283.68$$

$$\frac{-20w}{-20} = \frac{-4.8}{-20}$$

$$j = .68 \qquad w = .34$$

Answers
August 2016
Algebra I

Answer Key

PART I

1. (1)	**5.** (2)	**9.** (3)	**13.** (3)	**17.** (1)	**21.** (3)
2. (3)	**6.** (2)	**10.** (1)	**14.** (3)	**18.** (1)	**22.** (4)
3. (4)	**7.** (1)	**11.** (4)	**15.** (4)	**19.** (1)	**23.** (1)
4. (4)	**8.** (2)	**12.** (3)	**16.** (3)	**20.** (2)	**24.** (2)

PART II

25.

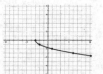

26. The graph of $h(x)$ is the graph of $b(x)$ translated 2 units to the right and 3 units down.

27. Exponential model

28. Pat is not correct. The leading exponent is –2.

29. Irrational

30. {–3, 1}

31. {–4, 10}

32. $\dfrac{13+3a}{a}$

PART III

33. $y = 17.159x - 2,476$, 8.7 degrees

34. The wrong side of the line is shaded.

35. 48 tickets

36. $x = -\dfrac{3}{8}$, $x = \dfrac{1}{8}$. She substituted $8x = B$.

PART IV

37. $18j + 32w = 19.92$,
$14j + 26w = 15.76$,
$w = 0.24$, $j = 0.68$

In Parts II–IV, you are required to show how you arrived at your answers. For sample methods of solutions, see the *Answers Explained* section.

Answers Explained

PART I

1. The rate of change on an interval is the change in miles divided by the change in time. The rate of change for each choice must be calculated and compared.

- Choice (1):

$$\text{rate of change} = \frac{2-0}{1-0}$$
$$= \frac{2}{1}$$
$$= 2$$

- Choice (2):

$$\text{rate of change} = \frac{3.5-2}{2-1}$$
$$= \frac{1.5}{1}$$
$$= 1.5$$

- Choice (3):

$$\text{rate of change} = \frac{4.5-3.5}{3-2}$$
$$= \frac{1}{1}$$
$$= 1$$

- Choice (4):

$$\text{rate of change} = \frac{5-4.5}{4-3}$$
$$= \frac{0.5}{1}$$
$$= 0.5$$

Choice (1) shows the greatest rate of change among the four choices, 2 miles per hour.

This question can also be answered without making any calculations by looking at the graph and checking if one line segment has a greater slope than the other line segments. Since the line segment connecting $(0, 0)$ and $(1, 2)$ is steeper than the other line segments, it corresponds to the interval with the greatest rate of change.

The correct choice is **(1)**.

2. A two-variable equation like $y = x + 3$ has an infinite number of solutions. Each solution is an ordered pair, an x-value and a y-value that make the equation true. For this equation, one ordered pair that makes it true is $(2, 5)$ since $5 = 2 + 3$. Another ordered pair that makes this equation true is $(4, 7)$ since $7 = 4 + 3$. This is what choice (3) is describing.

The correct choice is **(3)**.

3. A box plot describes five statistics about a data set: the minimum value, the maximum value, the median, the first quartile, and the third quartile. Here is an example of a box plot for a data set with a minimum of 60, a maximum of 100, a median of 82, a first quartile of 70, and a third quartile of 88.

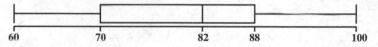

A box plot does not show the most frequent score.

The correct choice is **(4)**.

4. Since all the charts have x-values of 0, 2, 4, and 6, calculate $f(0), f(2), f(4)$, and $f(6)$. Then compare the values to the charts in the answer choices:

$$f(0) = -2(0) + 6$$
$$= 0 + 6$$
$$= 6$$
$$f(2) = -2(2) + 6$$
$$= -4 + 6$$
$$= 2$$

$$f(4) = -2(4) + 6$$
$$= -8 + 6$$
$$= -2$$

$$f(6) = -2(6) + 6$$
$$= -12 + 6$$
$$= -6$$

These values agree with the numbers in choice (4).

The correct choice is **(4)**.

5. Check each of the four choices by substituting the number inside the parentheses for all occurrences of n in the formula:

- Choice (1):

$$f(3) = (3 - 1)^2 + 3(3)$$
$$= 2^2 + 9$$
$$= 4 + 9$$
$$= 13 \neq -2$$

- Choice (2):

$$f(-2) = (-2 - 1)^2 + 3(-2)$$
$$= (-3)^2 - 6$$
$$= 9 - 6 = 3$$

- Choice (3):

$$f(-2) = (-2 - 1)^2 + 3(-2)$$
$$= (-3)^2 - 6$$
$$= 9 - 6$$
$$= 3 \neq -15$$

- Choice (4):

$$f(-15) = (-15 - 1)^2 + 3(-15)$$
$$= (-16)^2 - 45$$
$$= 256 - 45$$
$$= 211 \neq -2$$

The correct choice is **(2)**.

6. A dot plot for the data would look like this:

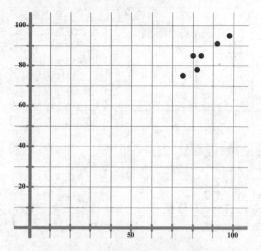

Since the scatter plot resembles a line with positive slope, the correlation coefficient would be very close to +1.

If the scatter plot resembled a line with negative slope, the correlation coefficient would be very close to −1.

If the scatter plot did not resemble a line at all but looked like a random cloud, the correlation coefficient would be close to 0.

If the scatter plot kind of resembled a line but many of the points were far from that line, it could have a correlation coefficient above 0 but not close to 1.

By using the graphing calculator, you can determine the exact value of the correlation coefficient. Put the numbers 92, 98, 84, 80, 75, and 82 into the first column (list L1) and the numbers 91, 95, 85, 85, 75, and 78 into the second column (list L2). Use the linear regression feature of the calculator to find that the r-value is approximately 0.92, which is very close to 1. For the TI-84, diagnostics must be turned on for the calculator to show the correlation coefficient. To turn diagnostics on, press [2ND] [0]. Scroll down to DiagnosticOn, and select it.

For the TI-84:

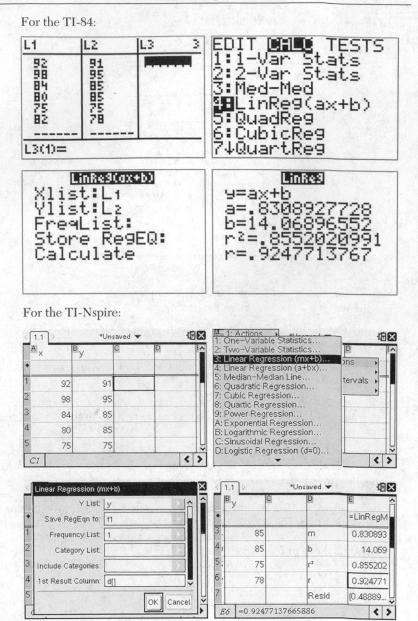

For the TI-Nspire:

The correct choice is (**2**).

7. Solving an inequality is almost the same as solving an equality. The only difference is that whenever you divide by a negative, you must reverse the direction of the inequality sign:

$$2h + 8 > 3h - 6$$
$$-3h = -3h$$
$$-h + 8 > -6$$
$$-8 = -8$$
$$-h > -14$$
$$\frac{h}{-1} < \frac{-14}{-1}$$
$$h < 14$$

The correct choice is (**1**).

8. Since the two terms in the expression have a common factor of 4, it can be factored out to make $4(9x^2 - 25)$.

Expressions of the form $a^2 - b^2$ are called the difference of perfect squares and can be factored into $(a - b)(a + b)$.

The expression $9x^2 - 25$ can be written as $(3x)^2 - 5^2$. So it matches the difference of perfect squares pattern and can be factored into $(3x - 5)(3x + 5)$.

So the original expression is equal to $4(3x - 5)(3x + 5)$. This is also equivalent to $4(3x + 5)(3x - 5)$.

Another way to do this problem would be to multiply each of the answer choices to see which one becomes $36x^2 - 100$.

For choice (2):

$$4(3x + 5)(3x - 5) = 4(9x^2 - 15x + 15x - 25)$$
$$= 4(9x^2 - 25)$$
$$= 36x^2 - 100$$

The correct choice is (**2**).

9. This question relies more on background knowledge and experience than actual math. A big rainstorm might bring a few inches of rain in a day, so feet per hour or inches per hour would not be appropriate. It is rare for even 1 foot of rain to fall in a month in New York, so feet per month would also not be appropriate.

The correct choice is (**3**).

10. After the first term, each term is 4 less than the previous term. After –18, the next two terms would be –22 and –26. When each term is the same amount more (or the same amount less) than the term before, you are working with an arithmetic sequence. The formula for the nth term of an arithmetic sequence is $f(x) = f(1) + d(x - 1)$, where $f(1)$ is the first term of the sequence and d is the common difference.

For this sequence, the first term is –6 and the common difference is –4 since each term is 4 less than the previous term. The function is therefore $f(x) = -6 + (-4)(x - 1)$. The function can be simplified to $f(x) = -6 - 4x + 4 = -4x - 2$.

A simpler way to solve this question, since it is multiple choice, would be to calculate $f(6)$ for each of the answer choices to see if any of them become –26. When testing choice (1), you would find

$$f(6) = -4(6) - 2 = -24 - 2 = -26$$

The correct choice is **(1)**.

11. Calculate the y-intercept for each of the four choices.

- Choice (1): The y-intercept is the y-coordinate when x equals 0. Since $f(0) = 3(0) = 0$, the y-intercept is 0.

- Choice (2): The y-intercept is the y-coordinate when x equals 0. Substitute 0 for x and solve for y:

$$2(0) + 3y = 12$$
$$0 + 3y = 12$$
$$3y = 12$$
$$\frac{3y}{3} = \frac{12}{3}$$
$$y = 4$$

The y-intercept for choice (2) is 4.

- Choice (3): The easiest way to find the y-intercept for this choice would be to make a sketch of the graph by plotting the point (1, –4) and making a line with a slope of 2. To find another point on the line, make it 1 unit to the right and 2 units up from (1, –4) to get to (2, –2). You can also find another point on the line by making it 1 unit to the left and 2 units down from (1, –4). This would get you to (0, –6). So the y-intercept is –6, which is definitely less than the y-intercept for either of the first two choices.

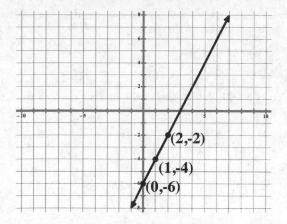

- Choice (4): On the graph, the line passes through the y-axis at the point (0, 5). So the y-intercept is 5, which is greater than the y-intercept for any of the other choices.

The correct choice is **(4)**.

12. To multiply these two polynomials, each of the two terms in the first polynomial must be multiplied by each of the three terms in the second polynomial. The resulting like terms must then be combined:

$$(2x + 3)(4x^2 - 5x + 6) = 2x \cdot 4x^2 + 2x(-5x) + 2x \cdot 6 + 3 \cdot 4x^2 + 3(-5x) + 3 \cdot 6$$
$$= 8x^3 - 10x^2 + 12x + 12x^2 - 15x + 18$$
$$= 8x^3 + 2x^2 - 3x + 18$$

The correct choice is **(3)**.

13. Analyze the four answer choices:

- Choice (1): At Time = 0, the rocket was at a height of 180 feet above the ground. So that was how high above the ground the rocket was when it was launched.

- Choice (2): At Time = 3, the rocket was 324 feet above the ground. This was the highest number for any of the times on the chart, so 324 feet was the maximum height.

- Choice (3): Since the rocket was still above the ground at Time = 7, it was in the air for at least 7 seconds, not 6.

- Choice (4): The rocket was 308 feet above the ground from Time = 2 until Time = 4, which is 2 seconds.

The correct choice is **(3)**.

14. From the chart, it can be seen that each additional hour after the first 2 hours costs $5.50. Solve each of the answer choices for x to see which equals 5.5:

- Choice (1):

$$9.00 + 3x = 20.00$$
$$-9.00 \qquad = -9.00$$
$$3x = 11.00$$
$$\frac{3x}{3} = \frac{11.00}{3}$$
$$x \approx 3.67$$

- Choice (2):

$$9.00 + 3.50x = 20.00$$
$$-9.00 \qquad = -9.00$$
$$3.50x = 11.00$$
$$\frac{3.50x}{3.50} = \frac{11.00}{3.50}$$
$$x \approx 3.14$$

- Choice (3):

$$2x + 3.50 = 14.50$$
$$-3.50 = -3.50$$
$$2x = 11$$
$$\frac{2x}{2} = \frac{11}{2}$$
$$x = 5.5$$

- Choice (4):

$$2x + 9.00 = 14.50$$
$$-9.00 = -9.00$$
$$2x = 5.50$$
$$\frac{2x}{2} = \frac{5.50}{2}$$
$$x = 2.75$$

The correct choice is **(3)**.

15. Examine the four choices.

    - Choice (1): Every time x increases by 1, y increases by 3. This means the rate of change is 3, not –3.

    - Choice (2): This function does not have a constant rate of change. When x increases from 1 to 2, y decreases by 3. So there is a rate of change of –3 for the interval between 1 and 2. For the interval between $x = 2$ and $x = 3$, though, the y-value changes by –7. So the rate of change for that interval is –7. Though there was one interval where the rate of change was –3, the question asks for the function that has a constant rate of change of –3.

    - Choice (3): For the graph of a linear function, the slope of the line is the constant rate of change. The slope of this line is –2, not –3.

    - Choice (4): The slope of the graph of this function can be found by first converting this equation into slope-intercept form and then looking at the coefficient of x:

    $$2y = -6x + 10$$
    $$\frac{2y}{2} = \frac{-6x + 10}{2}$$
    $$y = -3x + 5$$

    Since the coefficient of x is –3, the constant rate of change is –3.

    The correct choice is **(4)**.

16. Since Kendal kept 2 boxes, the number of boxes of cookies she brought to the party was $x - 2$. There are 12 cookies in each box. So if $x - 2$ boxes contain 60 cookies, the following equation can be used to solve for x:

    $$12(x - 2) = 60$$

    This is not one of the choices. By using the distributive property on the left side of the equal sign, though, it becomes the equivalent equation shown in choice (3):

    $$12x - 24 = 60$$

    Another way to solve this question would be to divide 60 by 12 to see that 5 boxes of cookies were brought to the party. Therefore, Kendal must have originally bought 7 boxes of cookies. Solve each of the choices for x. Only choice (3) becomes $x = 7$.

    The correct choice is **(3)**.

17. If you substitute $m = 0$ into each of the four choices, only choices (1) and (2) equal 150. So you can eliminate choices (3) and (4). If you then substitute $m = 2$ into choices (1) and (2), choice (1) equals 108.

    Another way to choose between choices (1) and (2) would be to notice that the numbers in the chart are decreasing. So the number being raised to the power must be less than 1.

    The correct choice is (**1**).

18. Substitute $x = 25$ into each of the four choices.

    - Choice (1):

    $$f(25) = 1.5^{25} \approx 25{,}251$$

    - Choice (2):

    $$g(25) = 1.5 \cdot 25 + 3 = 40.5$$

    - Choice (3):

    $$h(25) = 1.5 \cdot 25^2 = 937.5$$

    - Choice (4):

    $$k(25) = 1.5 \cdot 25^3 + 1.5 \cdot 25^2 = 24{,}375$$

    An exponential function will eventually be greater than a linear, quadratic, or cubic function for a large enough value of x. Still, you should check all answer choices using larger numbers, like $x = 100$, since 25 isn't such a large value of x. In fact when $x = 25$, choice (4) was nearly the biggest.

    The correct choice is (**1**).

19. This quadratic equation can be solved by factoring:

    $$3x^2 + 10x = 8$$
    $$-8 = -8$$
    $$3x^2 + 10x - 8 = 0$$
    $$(3x - 2)(x + 4) = 0$$
    $$3x - 2 = 0 \quad \text{or} \quad x + 4 = 0$$
    $$x = \frac{2}{3} \quad \text{or} \quad x = -4$$

    Another way to solve this question is to test the answer choices to see which has x-values that make the equation true.

- Test choice (1):

$$3\left(\frac{2}{3}\right)^2 + 10\left(\frac{2}{3}\right) \overset{?}{=} 8$$

$$\frac{12}{9} + \frac{20}{3} \overset{?}{=} 8$$

$$\frac{12}{9} + \frac{60}{9} \overset{?}{=} 8$$

$$\frac{72}{9} \overset{?}{=} 8$$

$$8 = 8 \quad \checkmark$$

Also:

$$3(-4)^2 + 10(-4) \overset{?}{=} 8$$

$$48 - 40 \overset{?}{=} 8$$

$$8 = 8 \quad \checkmark$$

The correct choice is **(1)**.

20. Since you cannot purchase a fraction of a song, the domain must be numbers like 1, 2, 3, 4, and so on. These are the whole numbers greater than or equal to one. If you could purchase a fraction of a song, then rational numbers greater than zero, like $\frac{1}{2}$, would be the domain.

The correct choice is **(2)**.

21. The vertex form of a quadratic equation is $a(x - h)^2 + k$, where the coordinates of the vertex are (h, k). The quickest way to find the correct choice is to find the vertex of the parabola that would be defined by the function $f(x) = 3x^2 + 12x + 11$.

First, find the x-coordinate of the vertex:

$$x = \frac{-b}{2a}$$

$$= \frac{-12}{2 \cdot 3}$$

$$= \frac{-12}{6}$$

$$= -2$$

To find the y-coordinate of the vertex, substitute the x-value you just found into the function:

$$f(-2) = 3(-2)^2 + 12(-2) + 11$$
$$= 12 - 24 + 11$$
$$= -1$$

The vertex is $(-2, -1)$. So the equation must be of the form

$$a(x - (-2))^2 + (-1) = a(x + 2)^2 - 1.$$

Only choice (3) has this form.

The correct choice is **(3)**.

22. Two systems of equations have the same solution if the equations in one are multiples of the equations in the other.

- Choice (1): The second equation of this system is the same as the second equation in the original system. The first equation in this system, $3x + 6y = 15$, is what you would get if you multiply both sides of the first equation of the original system by 3. So these systems will have the same solution.

- Choice (2): In this system, the second equation is the same as the second equation in the original system. The first equation in this system is equal to the first equation of the first system after both sides of that equation are multiplied by 4. So these systems have the same solution.

- Choice (3): In this system, the first equation matches the first equation from the original system. The second equation is equivalent to the second equation in the original system after both sides of that equation are multiplied by 3.

- Choice (4): The first equation matches but the second equation is not a multiple of the original second equation. It is what you would get if you multiplied the left side of the equation by 4 and the right side of the equation by 3.

An alternative, though much slower, way to answer this question would be to solve the original system of equations using the elimination method:

$$x + 2y = 5$$
$$2x + y = 4$$

To eliminate the y, multiply both sides of the second equation by -2, and solve for x:

$$x + 2y = 5$$
$$-2(2x + y) = -2(4)$$
$$x + 2y = 5$$
$$\underline{-4x - 2y = -8}$$
$$-3x = -3$$
$$x = 1$$

Then substitute $x = 1$ into one of the original equations to solve for y:

$$1 + 2y = 5$$
$$-1 = -1$$
$$2y = 4$$
$$y = 2$$

You could then solve the systems of equations in the answer choices to see which one does not have a solution of $(1, 2)$.

The correct choice is **(4)**.

23. The key to this question is to notice that the graph has three x-intercepts: $(-3, 0)$, $(2, 0)$, and $(4, 0)$. This means that if the numbers -3, 2, or 4 are substituted into the function for x, the function's value should be 0.

In general if a graph has an x-intercept of a, the equation will have a factor of $(x - a)$. Since this graph has the three x-intercepts -3, 2, and 4, it will have the factors $(x - (-3))$, $(x - 2)$, and $(x - 4)$. So the factorization could be $(x - (-3))(x - 2)(x - 4) = (x + 3)(x - 2)(x - 4)$.

The correct choice is **(1)**.

24. In an exponential function $a(1 + r)^x$, the r is the growth rate. When the number in the parentheses is less than 1, the value of r is negative and is called the exponential decay.

In this example the function can be written as $v(x) - 30{,}000(1 - 0.22)^x$. So the growth rate is -0.22, which is a 22% decrease per year.

The correct choice is **(2)**.

PART II

25. The graph of $y = \sqrt{x}$ contains the points $(0, 0)$, $(1, 1)$, $(4, 2)$, and $(9, 3)$ and looks like this.

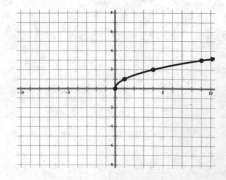

The graph of $y = -\sqrt{x+3}$ is like the graph of $y = \sqrt{x}$ but reflected over the x-axis and translated 3 units to the left. It contains the points $(-3, 0)$, $(-2, -1)$, $(1, -2)$, and $(6, -3)$.

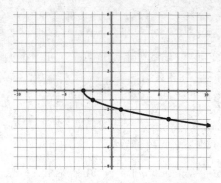

The graph can also be made using a graphing calculator.

On the TI-84:

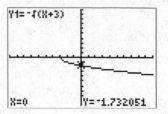

On the TI-Nspire:

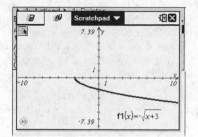

You can also make a table of x- and y-values using a graphing calculator, which can help you create the graph on your test booklet.

For the TI-84:

Press [2ND] [GRAPH].

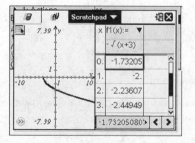

For the TI-Nspire:

Hold down [ctrl] and press [T].

26. When a function is defined in terms of another function, the graph of the new function is related to the graph of the original function. When a number is added (or subtracted) inside the parentheses of the original function, the graph of the new function is translated to the left (or right) by that number. When a number is added (or subtracted) outside the parentheses of the original function, the graph of the new function is translated up (or down) by that number.

In this case, $h(x) = b(x - 2) - 3$. So the graph of $h(x)$ will be like the graph of $b(x)$ but shifted 2 units to the right and 3 units down.

27. The sequence of the number of squares in each figure is 2, 4, 8, ... Each figure has double the number of squares as the previous figure. This is known as a geometric sequence. So the nth term would be 2^n. When the variable in an expression is the exponent, it is an exponential equation.

This pattern can be described with an exponential model.

28. Pat is not correct. Even though –4 is the first coefficient the way these terms are ordered, the leading coefficient is the number in front of the variable with the highest exponent. The highest exponent is 3. So the leading coefficient is –2.

Usually the terms of a polynomial appear so the exponents are arranged from greatest to least, like $-2x^3 + 8x^2 - 4x + 5$. When this happens, the leading coefficient is also the first coefficient.

So Pat is wrong. The leading coefficient is –2.

29. To add two radicals that have the same radical part, add the numbers in front of the radicals as you would with like terms:

$$3\sqrt{2} + 4\sqrt{2} = 7\sqrt{2}$$

This is an irrational number.

Had the question asked for the <u>product</u>, it would have been rational:

$$3\sqrt{2} \cdot 4\sqrt{2} = 12\sqrt{4}$$
$$= 12 \cdot 2$$
$$= 24$$

The sum of $3\sqrt{2}$ and $4\sqrt{2}$ is irrational.

30. The x-coordinates of the intersection point(s) of the two graphs are the values of x for which $f(x) = g(x)$. These graphs intersect at two points, $(-3, 4)$ and $(1, 3)$. So $f(-3) = g(-3) = 4$ and $f(1) = g(1) = 3$.

The values -3 and 1 are the two values for which $f(x) = g(x)$.

31. The zeros of a function are the values that make the function equal to zero. To find the zeros of this function, solve the equation $0 = (x - 3)^2 - 49$.

One way to solve this equation is to begin by adding 49 to both sides of the equation:

$$0 = (x - 3)^2 - 49$$
$$\underline{+49 = +49}$$
$$49 = (x - 3)^2$$
$$\pm\sqrt{49} = \sqrt{(x-3)^2}$$
$$\pm 7 = x - 3$$
$$\underline{+3 = +3}$$
$$3 \pm 7 = x$$
$$3 + 7 = x \text{ or } 3 - 7 = x$$
$$10 = x \text{ or } \quad -4 = x$$

Another way to solve this question is to simplify the right side and use factoring:

$$0 = (x - 3)^2 - 49$$
$$0 = x^2 - 6x + 9 - 49$$
$$0 = x^2 - 6x - 40$$
$$0 = (x - 10)(x + 4)$$
$$x - 10 = 0 \quad \text{or} \quad x + 4 = 0$$
$$x = 10 \quad \text{or} \qquad x = -4$$

32. To solve for x in terms of a, all the terms containing x must be moved to one side of the equation and all the terms without x must be moved to the other side of the equation:

$$4(ax + 3) - 3ax = 25 + 3a$$
$$4ax + 12 - 3ax = 25 + 3a$$
$$-12 = -12$$
$$4ax - 3ax = 13 + 3a$$

Factor the x out of the right side:

$$x(4a - 3a) = 13 + 3a$$
$$xa = 13 + 3a$$

Isolate the x by dividing both sides of the equation by a:

$$\frac{xa}{a} = \frac{13 + 3a}{a}$$
$$x = \frac{13 + 3a}{a}$$

PART III

33. The line of best fit can be found with a graphing calculator.

For the TI-84:

Press [STAT] [1]. Enter the x-values into L1 and the y-values into L2. Press [STAT] [RIGHT] [4] for the LinReg function.

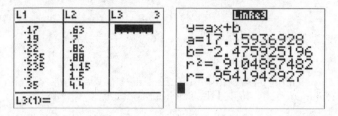

For the TI-Nspire:

From the home screen, go to "Lists & Spreadsheet." Enter the x-values into column A and label it "x." Enter the y-values into column B, and label it "y." Press [menu] [4] [1] [3] for "Linear Regression (mx+b)." Enter "x" for X-List and "y" for Y-List. Select the "OK" button.

The linear regression equation with the values rounded to the *nearest thousandth* is $y = 17.159x - 2.476$.

Substitute into this equation $x = 0.65$ to find the value of y:

$$y = 17.159 \cdot 0.65 - 2.476$$
$$= 8.67735$$
$$\approx 8.7$$

So the slope of the beach to the *nearest tenth of a degree* is approximately 8.7 degrees when the median diameter of the sand is 0.65 mm.

34. Shawn's mistake is that he shaded the wrong side of the line. The boundary line is properly graphed. It is a solid line because the inequality contains $\geq$ rather than $>$. The boundary line also correctly shows the equation $y = -\dfrac{1}{2}x - 4$. To decide which side of the line to shade, test to see if the ordered pair $(0, 0)$ makes the original inequality true.

For this example, $(0, 0)$ makes the inequality false:

$$-0 - 2 \cdot 0 \geq 8$$

$$0 \geq 8$$

The solution is not true.

So the shading needs to be on the side that does not contain $(0, 0)$. Shawn incorrectly shaded the side that contains $(0, 0)$.

The correct graph looks like this.

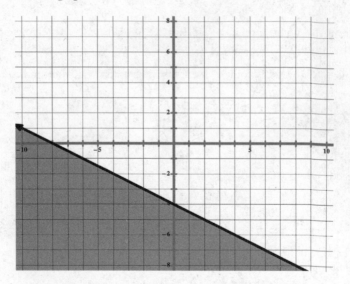

35. Let x equal the number of $12 tickets sold, and let y equal the number of $8.50 tickets sold. In order to sell at least $1000 worth of tickets, one inequality that must be true is the following:

$$12x + 8.5\,y \geq 1000$$

Since the auditorium holds 200 people, the total number of tickets sold must satisfy the following inequality:

$$x + y \leq 200$$

These two inequalities comprise a system of inequalities:

$$12x + 8.5y \geq 1000$$

$$x + y \leq 200$$

If 50 tickets are sold in advance, the amount of money collected for those tickets can be calculated:

$$\$8.50 \cdot 50 = \$425$$

To sell at least $1000 worth of tickets, the drama club must collect at least an additional $1000 − $425 = $575. The number of $12 tickets the drama club must sell to get $575 is $575 ÷ $12 ≈ 47.9. Since the club cannot sell a fraction of a ticket, round up. So a minimum of 48 tickets must be sold at the door.

36. Janice must have defined the variable $B = 8x$. She was then able to write the original equation as $0 = B^2 + 2B - 3$. After factoring, Janice could solve for two values of B that satisfy the new equation. Those B-values are $B = -3$ and $B = 1$.

To solve for x, replace the B with $8x$ and solve for two values of x.

$$8x = -3$$
$$x = -\frac{3}{8}$$

and

$$8x = 1$$
$$x = \frac{1}{8}$$

PART IV

37. If the juice boxes cost j dollars and the water bottles cost w dollars, the system of equations can be written as:

$$18j + 32w = 19.92$$

$$14j + 26w = 15.76$$

Kara cannot be correct. If you substitute $j = 0.52$ and $w = 0.33$ into the second equation, you get:

$$14 \cdot 0.52 + 26 \cdot 0.33 = 15.86 \neq 15.76$$

To find the actual answers, solve the system of equations. If you multiply both sides of the first equation by 14 and both sides of the second equation by -18, you can combine the two equations to eliminate j:

$$14(18j + 32w) = 14(19.92)$$

$$-18(14j + 26w) = -18(15.76)$$

$$252j + 448w = 278.88$$

$$\underline{-252j - 468w = -283.68}$$

$$-20w = -4.8$$

$$\frac{-20w}{-20} = \frac{-4.8}{-20}$$

$$w = 0.24$$

$$18j + 32(0.24) = 19.92$$

$$18j + 7.68 = 19.92$$

$$-7.68 = -7.68$$

$$18j = 12.24$$

$$\frac{18j}{18} = \frac{12.24}{18}$$

$$j = 0.68$$

So the true solution is that water bottles cost 24 cents each and juice boxes cost 68 cents each.

Topic	Question Numbers	Number of Points	Your Points	Your Percentage
1. Polynomials	12, 28	2 + 2 = 4		
2. Properties of Algebra	32	2		
3. Functions	1, 4, 5, 15, 18, 20, 26, 30	2 + 2 + 2 + 2 + 2 + 2 + 2 + 2 = 16		
4. Creating and Interpreting Equations	14, 16	2 + 2 = 4		
5. Inequalities	7, 34	2 + 4 = 6		
6. Sequences and Series	10	2		
7. Systems of Equations	22, 35, 37	2 + 4 + 6 = 12		
8. Quadratic Equations and Factoring	8, 13, 19, 21, 31, 36	2 + 2 + 2 + 2 + 2 + 4 = 14		
9. Regression	6, 33	2 + 4 = 6		
10. Exponential Equations	17, 24, 27	2 + 2 + 2 = 6		
11. Graphing	2, 11, 23, 25	2 + 2 + 2 + 2 = 8		
12. Statistics	3	2		
13. Number Properties	9, 29	2 + 2 = 4		

HOW TO CONVERT YOUR RAW SCORE TO YOUR ALGEBRA I REGENTS EXAMINATION SCORE

The accompanying conversion chart must be used to determine your final score on the August 2016 Regents Examination in Algebra I. To find your final exam score, locate in the column labeled "Raw Score" the total number of points you scored out of a possible 86 points. Since partial credit is allowed in Parts II, III, and IV of the test, you may need to approximate the credit you would receive for a solution that is not completely correct. Then locate in the adjacent column to the right the scale score that corresponds to your raw score. The scale score is your final Algebra I Regents Examination score.

Regents Examination in Algebra I—August 2016
Chart for Converting Total Test Raw Scores to Final
Examination Scores (Scaled Scores)

Raw Score	Scale Score	Performance Level	Raw Score	Scale Score	Performance Level	Raw Score	Scale Score	Performance Level
86	100	5	57	81	4	28	66	3
85	99	5	56	81	4	27	65	3
84	97	5	55	80	4	26	64	2
83	96	5	54	80	4	25	62	2
82	95	5	53	80	4	24	61	2
81	94	5	52	80	4	23	60	2
80	93	5	51	79	3	22	58	2
79	92	5	50	79	3	21	57	2
78	91	5	49	79	3	20	55	2
77	90	5	48	78	3	19	53	1
76	90	5	47	78	3	18	52	1
75	89	5	46	78	3	17	50	1
74	88	5	45	77	3	16	48	1
73	88	5	44	77	3	15	46	1
72	87	5	43	77	3	14	44	1
71	86	5	42	76	3	13	42	1
70	86	5	41	76	3	12	39	1
69	86	5	40	75	3	11	37	1
68	85	5	39	75	3	10	34	1
67	84	4	38	74	3	9	32	1
66	84	4	37	73	3	8	29	1
65	83	4	36	73	3	7	26	1
64	83	4	35	72	3	6	23	1
63	83	4	34	71	3	5	19	1
62	82	4	33	70	3	4	16	1
61	82	4	32	70	3	3	12	1
60	82	4	31	69	3	2	8	1
59	82	4	30	68	3	1	4	1
58	81	4	29	67	3	0	0	1

Examination
June 2017
Algebra I

HIGH SCHOOL MATH REFERENCE SHEET

Conversions

1 inch = 2.54 centimeters
1 meter = 39.37 inches
1 mile = 5280 feet
1 mile = 1760 yards
1 mile = 1.609 kilometers

1 kilometer = 0.62 mile
1 pound = 16 ounces
1 pound = 0.454 kilogram
1 kilogram = 2.2 pounds
1 ton = 2000 pounds

1 cup = 8 fluid ounces
1 pint = 2 cups
1 quart = 2 pints
1 gallon = 4 quarts
1 gallon = 3.785 liters
1 liter = 0.264 gallon
1 liter = 1000 cubic centimeters

Formulas

Triangle	$A = \dfrac{1}{2}bh$
Parallelogram	$A = bh$
Circle	$A = \pi r^2$
Circle	$C = \pi d$ or $C = 2\pi r$

Formulas (continued)

General Prisms	$V = Bh$
Cylinder	$V = \pi r^2 h$
Sphere	$V = \frac{4}{3}\pi r^3$
Cone	$V = \frac{1}{3}\pi r^2 h$
Pyramid	$V = \frac{1}{3}Bh$
Pythagorean Theorem	$a^2 + b^2 = c^2$
Quadratic Formula	$x = \dfrac{-b \pm \sqrt{b^2 - 4ac}}{2a}$
Arithmetic Sequence	$a_n = a_1 + (n - 1)d$
Geometric Sequence	$a_n = a_1 r^{n-1}$
Geometric Series	$S_n = \dfrac{a_1 - a_1 r^n}{1 - r}$ where $r \neq 1$
Radians	1 radian $= \dfrac{180}{\pi}$ degrees
Degrees	1 degree $= \dfrac{\pi}{180}$ radians
Exponential Growth/Decay	$A = A_0 e^{k(t - t_0)} + B_0$

PART I

Answer all 24 questions in this part. Each correct answer will receive 2 credits. No partial credit will be allowed. For each statement or question, write in the space provided the numeral preceding the word or expression that best completes the statement or answers the question. [48 credits]

1 To keep track of his profits, the owner of a carnival booth decided to model his ticket sales on a graph. He found that his profits only declined when he sold between 10 and 40 tickets. Which graph could represent his profits?

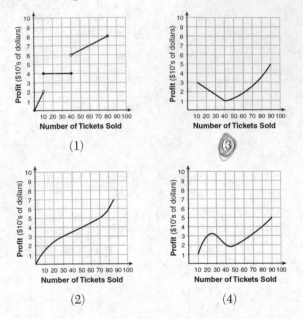

1 _____

2 The formula for the surface area of a right rectangular prism is $A = 2lw + 2hw + 2lh$, where l, w, and h represent the length, width, and height, respectively. Which term of this formula is *not* dependent on the height?

(1) A (3) $2hw$

(2) $2lw$ (4) $2lh$

2 _____

3 Which graph represents $y = \sqrt{x-2}$?

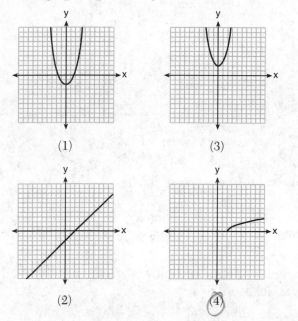

(1)   (3)

(2)   (4)

3 _____

4 A student plotted the data from a sleep study as shown in the graph below.

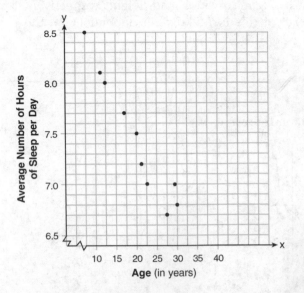

The student used the equation of the line

$$y = -0.09x + 9.24$$

to model the data. What does the rate of change represent in terms of these data?

(1) The average number of hours of sleep per day increases 0.09 hour per year of age.

(2) The average number of hours of sleep per day decreases 0.09 hour per year of age.

(3) The average number of hours of sleep per day increases 9.24 hours per year of age.

(4) The average number of hours of sleep per day decreases 9.24 hours per year of age.

4 _____

5 Lynn, Jude, and Anne were given the function $f(x) = -2x^2 + 32$, and they were asked to find $f(3)$. Lynn's answer was 14, Jude's answer was 4, and Anne's answer was ±4. Who is correct?

(1) Lynn, only   (3) Anne, only
(2) Jude, only   (4) Both Lynn and Jude   5 _____

6 Which expression is equivalent to $16x^4 - 64$?

(1) $(4x^2 - 8)^2$ (3) $(4x^2 + 8)(4x^2 - 8)$
(2) $(8x^2 - 32)^2$ (4) $(8x^2 + 32)(8x^2 - 32)$ 6 _____

7 Vinny collects population data, $P(h)$, about a specific strain of bacteria over time in hours, h, as shown in the graph below.

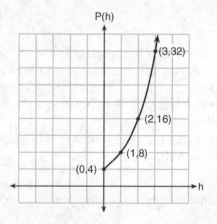

Which equation represents the graph of $P(h)$?

(1) $P(h) = 4(2)^h$

(2) $P(h) = \dfrac{16}{5}h + \dfrac{6}{5}$

(3) $P(h) - 3h^2 + 0.2h + 4.2$

(4) $P(h) = \dfrac{2}{3}h^3 - h^2 + 3h + 4$ 7 _____

8 What is the solution to the system of equations below?

$$y = 2x + 8$$
$$3(-2x + y) = 12$$

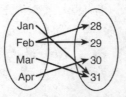

$3(-2x + (2x+8)=12$
$-6x+6x$

(1) no solution   (3) $(-1, 6)$

(2) infinite solutions   (4) $\left(\dfrac{1}{2}, 9\right)$

8 _____

9 A mapping is shown in the diagram below.

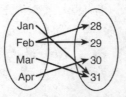

This mapping is

(1) a function, because Feb has two outputs, 28 and 29
(2) a function, because two inputs, Jan and Mar, result in the output 31
(3) not a function, because Feb has two outputs, 28 and 29
(4) not a function, because two inputs, Jan and Mar, result in the output 31

9 _____

10 Which polynomial function has zeros at −3, 0, and 4?

(1) $f(x) = (x + 3)(x^2 + 4)$
(2) $f(x) = (x^2 - 3)(x - 4)$
(3) $f(x) = x(x + 3)(x - 4)$
(4) $f(x) = x(x - 3)(x + 4)$

10 _____

11 Jordan works for a landscape company during his summer vacation. He is paid $12 per hour for mowing lawns and $14 per hour for planting gardens. He can work a maximum of 40 hours per week, and would like to earn at least $250 this week. If m represents the number of hours mowing lawns and g represents the number of hours planting gardens, which system of inequalities could be used to represent the given conditions?

(1) $m + g \leq 40$ (3) $m + g \leq 40$
     $12m + 14g \geq 250$ $12m + 14g \leq 250$

(2) $m + g \geq 40$ (4) $m + g \geq 40$
     $12m + 14g \leq 250$ $12m + 14g \geq 250$ 11 _____

12 Anne invested $1000 in an account with a 1.3% annual interest rate. She made no deposits or withdrawals on the account for 2 years. If interest was compounded annually, which equation represents the balance in the account after the 2 years?

(1) $A = 1000(1 - 0.013)^2$

(2) $A = 1000(1 + 0.013)^2$

(3) $A = 1000(1 - 1.3)^2$

(4) $A = 1000(1 + 1.3)^2$ 12 _____

13 Which value would be a solution for x in the inequality $47 - 4x < 7$?

(1) −13   (3) 10

(2) −10   (4) 11   13 _____

14 Bella recorded data and used her graphing calculator to find the equation for the line of best fit. She
then used the correlation coefficient to determine the
strength of the linear fit.

Which correlation coefficient represents the strongest
linear relationship?

(1) 0.9   (3) –0.3

(2) 0.5   (4) –0.8   14 _____

15 The heights, in inches, of 12 students are listed below.

   61, 67, 72, 62, 65, 59, 60, 79, 60, 61, 64, 63

Which statement best describes the spread of these
data?

(1) The set of data is evenly spread.

(2) The median of the data is 59.5.

(3) The set of data is skewed because 59 is the only
value below 60.

(4) 79 is an outlier, which would affect the standard
deviation of these data.   15 _____

16 The graph of a quadratic function is shown below.

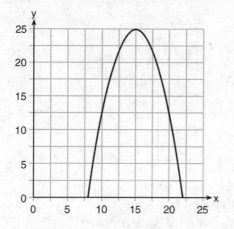

An equation that represents the function could be

(1) $q(x) = \dfrac{1}{2}(x + 15)^2 - 25$

(2) $q(x) = -\dfrac{1}{2}(x + 15)^2 - 25$

(3) $q(x) = \dfrac{1}{2}(x - 15)^2 + 25$

(4) $q(x) = -\dfrac{1}{2}(x - 15)^2 + 25$ 16 _____

17 Which statement is true about the quadratic functions $g(x)$, shown in the table below, and $f(x) = (x - 3)^2 + 2$?

x	g(x)
0	4
1	-1
2	-4
3	-5
4	-4
5	-1
6	4

(1) They have the same vertex.
(2) They have the same zeros.
(3) They have the same axis of symmetry.
(4) They intersect at two points.   17 _____

18 Given the function $f(n)$ defined by the following:

$$f(1) = 2$$
$$f(n) = -5f(n - 1) + 2$$

Which set could represent the range of the function?

(1) {2, 4, 6, 8,...}
(2) {2, -8, 42, -208,...}
(3) {-8, -42, -208, 1042,...}
(4) {-10, 50, -250, 1250,...}   18 _____

19 An equation is given below.

$$4(x - 7) = 0.3(x + 2) + 2.11$$

The solution to the equation is

(1) 8.3   (3) 3
(2) 8.7   (4) –3   19 ____

20 A construction worker needs to move 120 ft^3 of dirt by using a wheelbarrow. One wheelbarrow load holds 8 ft^3 of dirt and each load takes him 10 minutes to complete. One correct way to figure out the number of hours he would need to complete this job is

(1) $\dfrac{120 \text{ ft}^3}{1} \cdot \dfrac{10 \text{ min}}{1 \text{ load}} \cdot \dfrac{60 \text{ min}}{1 \text{ hr}} \cdot \dfrac{1 \text{ load}}{8 \text{ ft}^3}$

(2) $\dfrac{120 \text{ ft}^3}{1} \cdot \dfrac{60 \text{ min}}{1 \text{ hr}} \cdot \dfrac{8 \text{ ft}^3}{10 \text{ min}} \cdot \dfrac{1}{1 \text{ load}}$

(3) $\dfrac{120 \text{ ft}^3}{1} \cdot \dfrac{1 \text{ load}}{10 \text{ min}} \cdot \dfrac{8 \text{ ft}^3}{1 \text{ load}} \cdot \dfrac{1 \text{ hr}}{60 \text{ min}}$

(4) $\dfrac{120 \text{ ft}^3}{1} \cdot \dfrac{1 \text{ load}}{8 \text{ ft}^3} \cdot \dfrac{10 \text{ min}}{1 \text{ load}} \cdot \dfrac{1 \text{ hr}}{60 \text{ min}}$ 20 ____

21 One characteristic of all linear functions is that they change by

(1) equal factors over equal intervals
(2) unequal factors over equal intervals
(3) equal differences over equal intervals
(4) unequal differences over equal intervals   21 ____

22 What are the solutions to the equation $x^2 - 8x = 10$?

(1) $4 \pm \sqrt{10}$ (3) $-4 \pm \sqrt{10}$

(2) $4 \pm \sqrt{26}$ (4) $-4 \pm \sqrt{26}$ 22 _____

23 The formula for blood flow rate is given by $F = \dfrac{p_1 - p_2}{r}$, where F is the flow rate, p_1 the initial pressure, p_2 the final pressure, and r the resistance created by blood vessel size. Which formula can *not* be derived from the given formula?

(1) $p_1 = Fr + p_2$ (3) $r = F(p_2 - p_1)$

(2) $p_2 = p_1 - Fr$ (4) $r = \dfrac{p_1 - p_2}{F}$ 23 _____

24 Morgan throws a ball up into the air. The height of the ball above the ground, in feet, is modeled by the function $h(t) = -16t^2 + 24t$, where t represents the time, in seconds, since the ball was thrown. What is the appropriate domain for this situation?

(1) $0 \leq t \leq 1.5$ (3) $0 \leq h(t) \leq 1.5$

(2) $0 \leq t \leq 9$ (4) $0 \leq h(t) \leq 9$ 24 _____

PART II

Answer all 8 questions in this part. Each correct answer will receive 2 credits. Clearly indicate the necessary steps, including appropriate formula substitutions, diagrams, graphs, charts, etc. For all questions in this part, a correct numerical answer with no work shown will receive only 1 credit.   [16 credits]

25 Express in simplest form: $(3x^2 + 4x - 8) - (-2x^2 + 4x + 2)$

26 Graph the function $f(x) = -x^2 - 6x$ on the set of axes below.

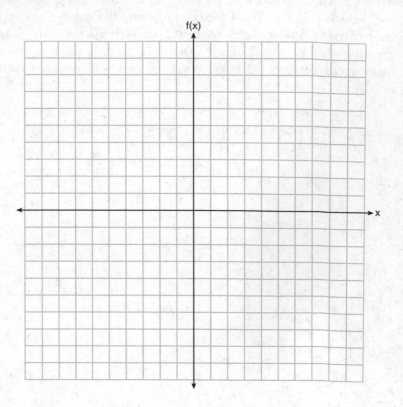

State the coordinates of the vertex of the graph.

27 State whether $7 - \sqrt{2}$ is rational or irrational. Explain your answer.

28 The value, $v(t)$, of a car depreciates according to the function $v(t) = P(.85)^t$, where P is the purchase price of the car and t is the time, in years, since the car was purchased. State the percent that the value of the car *decreases* by each year. Justify your answer.

29 A survey of 100 students was taken. It was found that 60 students watched sports, and 34 of these students did not like pop music. Of the students who did *not* watch sports, 70% liked pop music. Complete the two-way frequency table.

	Watch Sports	Don't Watch Sports	Total
Like Pop			
Don't Like Pop			
Total			

30 Graph the inequality $y + 4 < -2(x - 4)$ on the set of axes below.

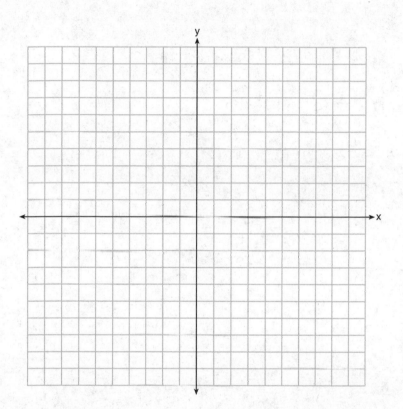

31 If $f(x) = x^2$ and $g(x) = x$, determine the value(s) of x that satisfy the equation $f(x) = g(x)$.

32 Describe the effect that each transformation below has on the function $f(x) = |x|$, where $a > 0$.

$g(x) = |x - a|$

$h(x) = |x| - a$

PART III

Answer all 4 questions in this part. Each correct answer will receive 4 credits. Clearly indicate the necessary steps, including appropriate formula substitutions, diagrams, graphs, charts, etc. For all questions in this part, a correct numerical answer with no work shown will receive only 1 credit.   [16 credits]

33 The function $r(x)$ is defined by the expression $x^2 + 3x - 18$. Use factoring to determine the zeros of $r(x)$.

Explain what the zeros represent on the graph of $r(x)$.

34 The graph below models Craig's trip to visit his friend in another state. In the course of his travels, he encountered both highway and city driving.

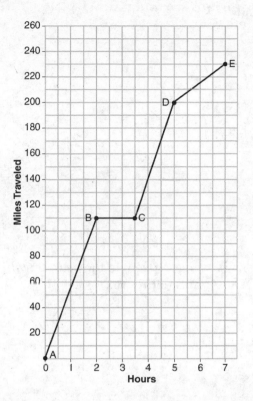

Based on the graph, during which interval did Craig most likely drive in the city? Explain your reasoning.

Question 34 is continued on the next page.

Question 34 continued.

Explain what might have happened in the interval between B and C.

Determine Craig's average speed, to the *nearest tenth of a mile per hour*, for his entire trip.

35 Given:

$$g(x) = 2x^2 + 3x + 10$$
$$k(x) = 2x + 16$$

Solve the equation $g(x) = 2k(x)$ algebraically for x, to the *nearest tenth*.

Explain why you chose the method you used to solve this quadratic equation.

36 Michael has $10 in his savings account. Option 1 will add $100 to his account each week. Option 2 will double the amount in his account at the end of each week.

Write a function in terms of x to model each option of saving.

Michael wants to have at least $700 in his account at the end of 7 weeks to buy a mountain bike. Determine which option(s) will enable him to reach his goal. Justify your answer.

PART IV

Answer the question in this part. A correct answer will receive 6 credits. Clearly indicate the necessary steps, including appropriate formula substitutions, diagrams, graphs, charts, etc. A correct numerical answer with no work shown will receive only 1 credit.   [6 credits]

37 Central High School had five members on their swim team in 2010. Over the next several years, the team increased by an average of 10 members per year. The same school had 35 members in their chorus in 2010. The chorus saw an increase of 5 members per year.

Write a system of equations to model this situation, where x represents the number of years since 2010.

Question 37 is continued on the next page.

Question 37 continued.

Graph this system of equations on the set of axes below.

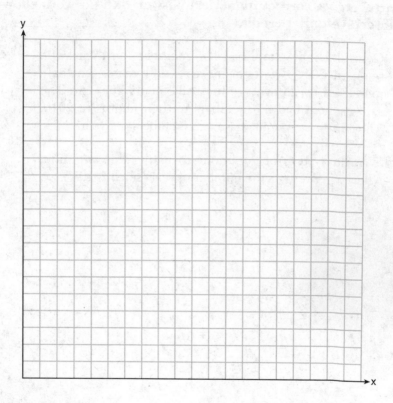

Explain in detail what each coordinate of the point of intersection of these equations means in the context of this problem.

Answers
June 2017
Algebra I

Answer Key

PART I

1. (3)	**5.** (1)	**9.** (3)	**13.** (4)	**17.** (3)	**21.** (3)
2. (2)	**6.** (3)	**10.** (3)	**14.** (1)	**18.** (2)	**22.** (2)
3. (4)	**7.** (1)	**11.** (1)	**15.** (4)	**19.** (1)	**23.** (3)
4. (2)	**8.** (1)	**12.** (2)	**16.** (4)	**20.** (4)	**24.** (1)

PART II

25. $5x^2 - 10$

26. Vertex: $(-3, 9)$

27. Irrational

28. 15%

29.

	Watch Sports	Don't Watch Sports	Total
Like Pop	26	28	54
Don't Like Pop	34	12	46
Total	60	40	100

30. See graph

31. $\{0, 1\}$

32. $g(x)$ is shifted a units to the right. $h(x)$ is shifted a units down.

PART III

33. The zeros are -6 and 3. They represent the x-intercepts of the graph.

34. Between D and E, the speed is 15 miles per hour, which is reasonable for city driving. Between B and C, he was stopped. The average speed for the entire trip is 32.9 miles per hour.

35. $\{3.6, -3.1\}$ The quadratic formula is simplest.

36. Option 1: $y = 100x + 10$. Option 2: $y = 10 \cdot 2^x$. Both are over $700 after 7 weeks.

PART IV

37. Swim team: $y = 10x + 5$. Chorus: $y = 5x + 35$. The intersection point is $(6, 65)$. In 2016, they both have 65 members.

In Parts II–IV, you are required to show how you arrived at your answers. For sample methods of solutions, see the *Answers Explained* section.

Answers Explained

PART I

1. If the profits are declining only when he sells between 10 and 40 tickets, the graph should show a curve that is going down as it goes from left to right between 10 and 40.

    The graph in choice (1) is flat between 10 and 40. The graph in choice (2) is going up between 10 and 40. The graph in choice (4) is going up from 10 to 25 but then down from 25 to 40. Only the graph in choice (3) is going down for the entire interval between 10 and 40.

    The correct choice is **(3)**.

2. A term is dependent on a variable if a change in that variable's value would also change the value of that entire term. When a term contains a variable, the term is dependent on that variable.

    Choices (3) and (4) have the variable h in them, so those can be eliminated. Even though choice (1) doesn't look like it has an h in it, it does. If you look at the definition of A, its equation has two hs in it. So A is also dependent on h. Choice (2) has no h in it or in the definition of any of the variables involved in that choice.

    The correct choice is **(2)**.

3. Graphing the relation $y = \sqrt{x-2}$ on the graphing calculator makes a graph that resembles choice **(4)**.

    For the TI-84:

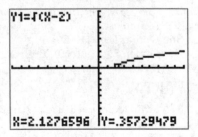

For the TI-Nspire:

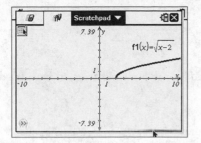

This graph can also be made without the graphing calculator by knowing that the graph of the function $y = \sqrt{x}$ has the following shape.

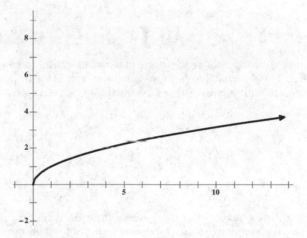

The graph of $y = \sqrt{x-2}$ is a transformation of the graph of $y = \sqrt{x}$ shifted 2 units to the right.

The correct choice is **(4)**.

4. The rate of change is equivalent to the slope of the line. Since m represents the slope when the equation is of the form $y = mx + b$, the rate of change for this model is -0.09. For every 1 unit that a point on the line goes to the right, the line will go down by 0.09 units.

For this model, a change by 1 unit to the right corresponds to an increase in age by 1 year. A change of 0.09 units down corresponds to a 0.09 decrease in the average number of hours of sleep per day.

The correct choice is **(2)**.

5. To calculate $f(3)$, substitute 3 for x in the equation:

$$f(3) = -2 \cdot 3^2 + 32$$
$$= -2 \cdot 9 + 32$$
$$= -18 + 32$$
$$= 14$$

Lynn got 14 for her answer. The other two people seem to have tried to solve the equation $0 = -2x^2 + 32$.

The correct choice is **(1)**.

6. Since both $16x^4$ and 64 are perfect squares, this expression can be written as $(4x^2)^2 - 8^2$. This can be factored as the difference of perfect squares:

$$a^2 - b^2 = (a - b)(a + b)$$
$$(4x^2)^2 - 8^2 = (4x^2 - 8)(4x^2 + 8)$$

The correct choice is **(3)**.

7. Each time the x-coordinate increases by 1 in this graph, the y-coordinate doubles. This is what happens when the graph is of an exponential equation, and only choice (1) is exponential.

Another way to solve this question is to calculate to see if $P(0) = 4$ and $P(1) = 8$ for any of the answer choices. Test choice (1):

$$P(0) = 4(2)^0 = 4 \cdot 1 = 4$$
$$P(1) = 4(2)^1 = 4 \cdot 2 = 8$$

The correct choice is **(1)**.

8. Since the first equation has the y-variable isolated on one side of the equation, the simplest way to solve this system of equations is with the substitution method. Substitute $2x + 8$ for the y in the second equation:

$$3(-2x + (2x + 8)) = 12$$
$$3(-2x + 2x + 8) = 12$$
$$3(8) = 12$$
$$24 = 12$$

Since 24 can never equal 12, this system of equations has no solution.

The correct choice is **(1)**.

9. In a function, each input value can be mapped to only one output value. When that happens, you can make statements like $f(\text{Jan}) = 31$ and $f(\text{Apr}) = 30$. In this mapping diagram though, Feb is mapped to two different numbers. So this mapping is not considered a function.

The correct choice is **(3)**.

10. A zero of a polynomial is a number that when put into the function makes an output of zero. For -3 to be a zero, $(x + 3)$ must be a factor since $-3 + 3 = 0$. For 4 to be a zero, $(x - 4)$ must be a factor. For 0 to be a zero, x must be a factor.

So the function is $f(x) = x(x + 3)(x - 4)$.

The correct choice is **(3)**.

11. Since Jordan can work at most 40 hours, the sum of his hours mowing and planting gardens must be less than or equal to 40. If he mows for m hours at \$12 per hour, the amount he makes mowing is $12m$. If he plants gardens for g hours at \$14 per hour, the amount he makes planting gardens is $14g$. Together he wants these amounts to be greater than or equal to 250.

So the system of inequalities is

$$m + \quad g \le 40$$
$$12m + 14g \ge 250$$

The correct choice is **(1)**.

12. After the first deposit, Anne has \$1000 in the account. After 1 year, she earns 1.3% interest. Her new balance can be calculated by multiplying \$1000 by 0.013 to find the amount of interest. Then add the interest to the original \$1000 to get $\$1000 + \$13 = \$1013$. To find the amount Anne has after 2 years, multiply \$1013 by 0.013 to find the amount of interest. Then add the interest to the \$1013 to get $\$1013 + \$13.169 = \$1026.169$. If you test each of the choices, only choice (2) gets this same answer.

A shorter way is to know that interest compounded annually can be calculated with the formula $A = P(1 + r)^n$, where P is the initial deposit, r is the percent interest as a decimal, and n is the number of years. In this example, $P = 1000$, $r = 0.013$, and $n = 2$.

The correct choice is **(2)**.

13. Start solving the inequality the same way you would solve an equation that contained an equal sign instead of a < sign:

$$47 - 4x < 7$$
$$-47 \quad\ = -47$$
$$-4x < -40$$

When you divide both sides of an inequality by a negative, you must flip the direction of the inequality sign:

$$\frac{-4x}{-4} > \frac{-40}{-4}$$
$$x > 10$$

Of the choices, only the number 11 is greater than 10.

You could answer this question by just testing each of the answer choices to see which of them make the inequality true. Test choice (4):

$$47 - 4(11) < 7$$
$$47 - 44 < 7$$
$$3 < 7$$

This is true. Testing the other choices would result in an inequality that is false.

The correct choice is **(4)**.

14. The closer the correlation coefficient is to either 1 or –1, the stronger the linear relationship is. Since 0.9 is just 0.1 away from 1 while –0.8 is 0.2 away from –1, 0.9 is the closest.

The correct choice is **(1)**.

15. Arrange the numbers from smallest to largest:

59, 60, 60, 61, 61, 62, 63, 64, 65, 67, 72, 79

Analyze each of the answer choices.

- Choice (1): Evenly spaced numbers would be something like 59, 61, 63, 65, 67, where each number is the same amount more than the number before. That is not what is happening with this set of numbers.

- Choice (2): The median is the middle number. Since there are an even amount of numbers, there is no one middle number. For an even amount of numbers, the median is the average of the two middle numbers, in this case the 6th and 7th numbers:

$$\text{median} = \frac{62+63}{2} = 62.5$$

The median is 62.5, not 59.5. Note that 59.5 is what you would get if you did not arrange the numbers in order before taking the average of the 6th and 7th numbers.

- Choice (3): Most data would be considered skewed. In this case, the data are not skewed because 59 is the only number less than 60.

- Choice (4): The standard deviation is a measure of how close together the numbers are. If all the numbers were the same, the standard deviation would be zero. Without the 79, the standard deviation would be much lower since all the other numbers are pretty close to 65.

For the TI-84:

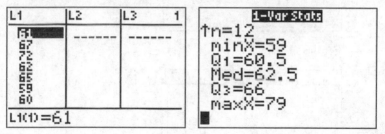

For the TI-Nspire:

The correct choice is (4).

16. The graph shows a parabola with a vertex at (15, 25) and that opens downward. The vertex form for the equation of a parabola is $y = a(x - h)^2 + k$, where h is the x-coordinate of the vertex, k is the y-coordinate of the vertex, a will be positive if the parabola opens upward, and a will be negative if the parabola opens downward. The graphs for choices (3) and (4) are parabolas with a vertex at (15, 25). Since choice (4) has a negative a-value, the parabola will open downward like in the question.

On the graphing calculator, the four choices can be graphed. Only choice (4) will resemble the graph in the question.

For the TI-84:

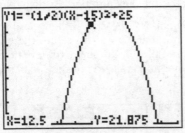

For the TI-Nspire:

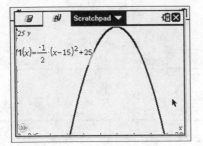

The correct choice is **(4)**.

17. The graphs of $f(x)$ and $g(x)$ look like this.

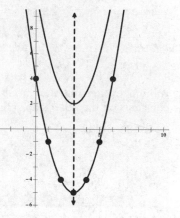

Analyze the answer choices:

• Choice (1): The vertex of $g(x)$ is $(3, -5)$, while the vertex of $f(x)$ is $(3, 2)$.

• Choice (2): Zeros can be seen on a graph as x intercepts. Function $g(x)$ has two zeros. One is between 0 and 1, and the other is between 5 and 6. Function $f(x)$ does not have any zeros.

• Choice (3): The axis of symmetry of $g(x)$ is $x = 3$, and the axis of symmetry of $f(x)$ is also $x = 3$.

• Choice (4): The graphs do not intersect.

The correct choice is **(3)**.

18. Since $f(1) = 2$, the first number in the sequence must be 2. Eliminate choices (3) and (4). To calculate the second number in the sequence, substitute 2 for n in the recursive part of the definition:

$$
\begin{aligned}
f(2) &= -5f(2-1) + 2 \\
&= -5f(1) + 2 \\
&= -5 \cdot 2 + 2 \\
&= -10 + 2 = -8 \\
&= -8
\end{aligned}
$$

The correct choice is **(2)**.

19. First use the distributive property:

$$4(x - 7) = 0.3(x + 2) + 2.11$$
$$4x - 28 = 0.3x + 0.6 + 2.11$$

Then combine like terms:

$$4x - 28 = 0.3x + 2.71$$

Complete the problem by getting the variables on one side of the equation and the constants on the other side of the equation:

$$4x - 28 = 0.3x + 2.71$$
$$-0.3x = -0.3x$$
$$3.7x - 28 = 2.71$$
$$+28 = +28$$
$$\frac{3.7x}{3.7} = \frac{30.71}{3.7}$$
$$x = 8.3$$

The correct choice is (**1**).

20. The simplest way to solve this is to calculate the number of hours to complete the job. To calculate the number of wheelbarrow loads needed, divide 120 by 8 to get 15 wheelbarrows. To calculate the number of minutes needed to move the dirt, multiply the number of wheelbarrow loads by 10, which is $15 \times 10 = 150$ minutes. Finally, convert 150 minutes to hours by dividing 150 by 60 to get 2.5 hours.

Test the four answer choices to see which one equals 2.5 hours.

The correct choice is (**4**).

21. A linear function is something like $f(x) = 2x^1 - 3 = 2x - 3$. In other words, the exponent on the variable is 1. The graph of a linear function is a line.

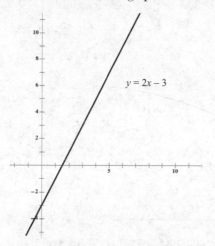

Start from any point on the line, such as $(2, 1)$, and pick an interval, such as 3. The point 3 to the right has the coordinates $(5, 7)$. The interval in this case is 3. The difference in the y-coordinates is $7 - 1 = 6$. If you start from some different point, such as $(4, 5)$, and choose the same interval, 3, the other point will be $(7, 11)$. The difference in the y-coordinates is 6. The differences are the same.

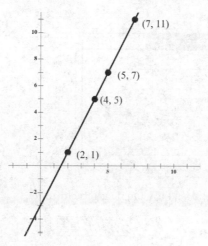

The correct choice is **(3)**.

22. There are three ways to solve this question.

One way is to complete the square by adding $\left(\frac{-8}{2}\right)^2 = (-4)^2 = 16$ to both sides of the equation:

$$x^2 - 8x = 10$$
$$+16 = +16$$
$$x^2 - 8x + 16 = 26$$
$$(x - 4)^2 = 26$$
$$\sqrt{(x-4)^2} = \pm\sqrt{26}$$
$$x - 4 = \pm\sqrt{26}$$

Another way to answer this is to substitute each of the answers for x into the expression $x^2 - 8x$ to see which one makes the expression evaluate to 16.

For the TI-84:

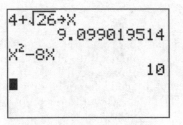

For the TI-Nspire:

Scratchpad ▼	
$4+\sqrt{26} \to x$	9.09902
$x^2 - 8 \cdot x$	10.
	2/99

The third method is to use the quadratic formula:

$$x = \frac{-b \pm \sqrt{b^2 - 4ac}}{2a}$$

Remember to start by rearranging the given equation into the format $ax^2 + bx + c = 0$:

$$x^2 - 8x = 10$$
$$x^2 - 8x - 10 = 0$$

$$x = \frac{-(-8) \pm \sqrt{(-8)^2 - 4(1)(-10)}}{(2)(1)}$$

$$x = \frac{8 \pm \sqrt{64 + 40}}{2}$$

$$x = \frac{8 \pm \sqrt{104}}{2}$$

$$x = \frac{8 \pm \sqrt{4 \cdot 26}}{2}$$

$$x = \frac{8 \pm 2\sqrt{26}}{2}$$

$$x = 4 \pm \sqrt{26}$$

The correct choice is **(2)**.

23. To isolate the other variables, start by multiplying both sides of the equation by r:

$$Fr = r \cdot \frac{p_1 - p_2}{r}$$
$$Fr = p_1 - p_2$$

The isolate r:

$$\frac{Fr}{F} = \frac{p_1 - p_2}{F}$$

This is choice (4).

Then isolate p_1:

$$Fr = p_1 - p_2$$
$$+p_2 = +p_2$$
$$Fr + p_2 = p_1$$

This is choice (1).

Then isolate p_2:

$$Fr = p_1 - p_2$$
$$+p_2 = + p_2$$
$$Fr + p_2 = p_1$$
$$-Fr = -Fr$$
$$p_2 = p_1 - Fr$$

This is choice (2).

Only choice (3) cannot be derived from the given formula.

The correct choice is **(3)**.

24. Since the ball never goes underground, the appropriate domain is the values of t of which $h(t) \geq 0$. By graphing the function on a graphing calculator, it can be seen that the ball starts on the ground at $t = 0$ and lands when $t = 1.5$.

For the TI-84:

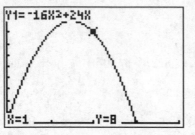

For the TI-Nspire:

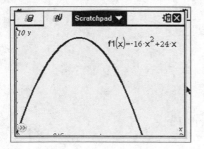

This can also be solved using algebra by calculating the values of t that make $h(t) = 0$:

$$-16t^2 + 24t = 0$$
$$8t(-2t + 3) = 0$$
$$8t = 0 \quad \text{or} \quad -2t + 3 = 0$$
$$t = 0 \quad \text{or} \quad -2t = -3$$
$$t = \frac{3}{2}$$

If the ball is on the ground at $t = 0$ and at $t = \frac{3}{2} = 1.5$, then it is above the ground between those two values.

The correct choice is **(1)**.

PART II

25. Eliminate the parentheses by distributing the negative sign through the second expression. Then combine like terms:

$$(3x^2 + 4x - 8) - (2x^2 + 4x + 2)$$
$$3x^2 + 4x - 8 - 2x^2 + 4x - 2$$
$$5x^2 - 10$$

26. The graph can be created with a table.

x	$f(x) = -x^2 - 6x$
-6	$-(-6)^2 - 6(-6) = -36 + 36 = 0$
-5	$-(-5)^2 - 6(-5) = -25 + 30 = 5$
-4	$-(-4)^2 - 6(-4) = -16 + 24 = 8$
-3	$-(-3)^2 - 6(-3) = -9 + 18 = 9$
-2	$-(-2)^2 - 6(-2) = -4 + 12 = 8$
-1	$-(-1)^2 - 6(-1) = -1 + 6 = 5$
0	$-(0)^2 - 6(0) = 0 + 0 = 0$

This table can also be created with the graphing calculator.

For the TI-84:

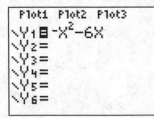

For the TI-Nspire:

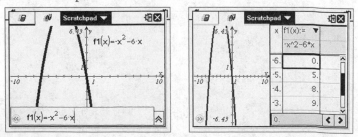

The graph looks like this:

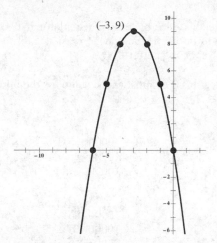

The vertex is the highest point of the parabola (or the lowest if it is a "smiling" parabola). The coordinates of the vertex, according to the graph, are (–3, 9).

You can also calculate the x-coordinate of the vertex by using the following formula:

$$x = \frac{-b}{2a}$$

$$= \frac{-(-6)}{2(-1)}$$

$$= \frac{6}{-2}$$

$$= -3$$

The y-coordinate of the vertex is what you get when you substitute the x-coordinate for x in the function:

$$y = f(-3)$$
$$= -(-3)^2 - 6(-3)$$
$$= -9 + 18$$
$$= 9$$

So the coordinates of the vertex are (–3, 9).

27. When an irrational number is added to or subtracted from a rational number, the result will be irrational.

    If you calculate this difference on a calculator it becomes $5.585786\ldots$, which appears to be irrational since it does not terminate and does not seem to have a repeating pattern.

28. One way to solve this is to substitute a value for the initial price of the car, like $P = 10000$.

    For $t = 0$:

    $$v(0) = 10000(.85)^0$$
    $$= 10000 \cdot 1$$
    $$= 10000$$

    One year later for $t = 1$:

    $$v(1) = 10000(.85)^1$$
    $$= 10000 \cdot .85$$
    $$= 8500$$

    Calculate the percent change from $t = 0$ to $t = 1$:

    $$\% \text{ change} = \frac{\text{old price} - \text{new price}}{\text{old price}}$$

    $$= \frac{10000 - 8500}{10000}$$

    $$= \frac{15000}{1000}$$

    $$= 0.15$$
    $$= 15\%$$

    A quicker way to solve this is to write the equation in the form $v(t) = P(1 + r)^t$, where r is the percent change. Since $.85 = (1 - .15)$, the percent change is 15%. Since r is negative, this is a percent decrease.

29. From the given information, three values from the chart can be quickly filled in.

	Watch Sports	Don't Watch Sports	Total
Like Pop			
Don't Like Pop	34		
Total	60		100

From these three values, the number of students who do not watch sports can be calculated by subtracting the number of students who watch sports from the total number of students:

$$100 - 60 = 40$$

	Watch Sports	Don't Watch Sports	Total
Like Pop			
Don't Like Pop	34		
Total	60	40	100

Since 70% of the students who do not watch sports do like pop music, calculate 70% of 40 and put that number into the table:

$$40 \cdot 0.7 = 28$$

	Watch Sports	Don't Watch Sports	Total
Like Pop		28	
Don't Like Pop	34		
Total	60	40	100

Now there is enough information to calculate the remaining four cells in the table. Each row and each column must add up to the total:

$$60 - 34 = 26$$
$$40 - 28 = 12$$
$$34 + 12 = 46$$
$$26 + 28 = 54$$

	Watch Sports	Don't Watch Sports	Total
Like Pop	26	28	54
Don't Like Pop	34	12	46
Total	60	40	100

30. First rewrite the inequality in slope-intercept form:

$$y + 4 < -2(x - 4)$$
$$y + 4 < -2x + 8$$
$$-4 = -4$$
$$y < -2x + 4$$

Graph the boundary line $y = -2x + 4$. Make it a dotted line because the inequality sign is < and not ≤. It has a slope of –2 and a y-intercept of 4.

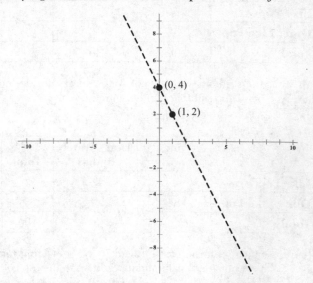

To decide which side of the line to shade, test to see if the point $(0, 0)$ makes the inequality true:

$$0 < -2(0) + 4$$
$$0 < 0 + 4$$
$$0 < 4$$

Since this is true, shade the side of the line containing (0, 0).

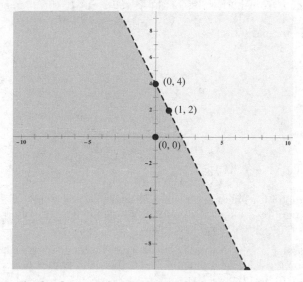

This can also be done on the graphing calculator.

For the TI-84:

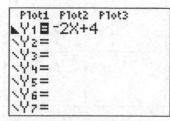
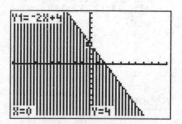

For the TI-Nspire:

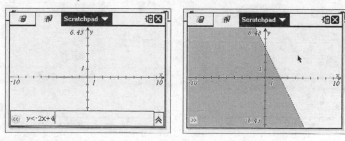

31. If $f(x) = g(x)$, then $x^2 = x$. To solve this equation, the first step is to subtract x from both sides of the equal sign. If you instead start by dividing both sides by x, you will lose one of the correct answers. Then factor and solve for x.

$$x^2 = x$$
$$-x = -x$$
$$x^2 - x = 0$$
$$x(x - 1) = 0$$

$$x = 0 \quad \text{or} \quad x - 1 = 0$$
$$+1 = +1$$
$$x = 1$$

The solution set is $\{0, 1\}$.

32. The graph of a function $f(x - a)$ will look like the graph of $f(x)$ shifted to the right by a units. The graph of a function $f(x) - a$ will look like the graph of $f(x)$ shifted down by a units.

By choosing a value for a, like $a = 2$, and graphing $f(x)$, $g(x)$, and $h(x)$ on a graphing calculator it is easier to see how the graphs of $g(x)$ and $h(x)$ compare to the graph of $f(x)$.

For the TI-84:

For the TI-Nspire:

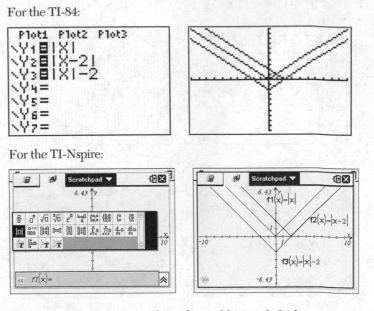

So $g(x)$ is shifted a units to the right, and $h(x)$ is shifted a units down.

PART III

33. The zeros of a function are the input values that make the function equal to 0. To find zeros, solve the equation:

$$r(x) = 0$$
$$x^2 + 3x - 18 = 0$$

Factor a polynomial in the form $x^2 + bx + c$ into $(x + p)(x + q)$ by finding, if possible, two numbers that have a sum of b and a product of c. In this case since b is 3 and c is -18, the two numbers are 6 and -3 since $6 + -3 = 3$ and $6 \cdot -3 = 18$:

$$x^2 + 3x - 18 = 0$$
$$(x + 6)(x - 3) = 0$$

$$x + 6 = 0 \quad \text{or} \quad x - 3 = 0$$
$$-6 = -6 \qquad\qquad +3 = +3$$
$$x = -6 \qquad\qquad x = 3$$

On the graph of the function, the zeros are the x-intercepts. The x-intercepts of the graph of $r(x)$ are $(-6, 0)$ and $(3, 0)$. This can be seen on the graphing calculator.

For the TI-84:

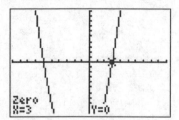

For the TI-Nspire:

34. In a graph where the x-axis is time and the y-axis is distance, the slope of a line segment is the speed at which the object is moving. There are four line segments in this graph. Find the slope of each segment.

- Calculate the slope from A to B.

  From hours = 0 to hours = 2, the distance changed from 0 miles to 110 miles:

$$\text{speed} = \text{slope}$$
$$= \frac{110 - 0}{2 - 0}$$
$$= \frac{110}{2}$$
$$= 55 \text{ miles per hour}$$

- Calculate the slope from B to C.

  From hours = 2 to hours = 3.5, the distance stayed at 110 for the whole time:

$$\text{speed} = \text{slope}$$
$$= \frac{110 - 110}{3.5 - 2}$$
$$= \frac{0}{1.5}$$
$$= 0 \text{ miles per hour (stopped)}$$

- Calculate the slope from C to D.

  From hours = 3.5 to hours = 5, the distance changed from 110 miles to 200 miles:

$$\text{speed} = \text{slope}$$
$$= \frac{200 - 110}{5 - 3.5}$$
$$= \frac{90}{1.5}$$
$$= 60 \text{ miles per hour}$$

- Calculate the slope from D to E.

From hours = 5 to hours = 7, the distance changed from 200 miles to 230 miles:

$$speed = slope$$

$$= \frac{230 - 200}{7 - 5}$$

$$= \frac{30}{2}$$

$$= 15 \text{ miles per hour}$$

Craig most likely drove in the city between D and E because city driving is most likely to be 15 miles per hour.

In the interval between B and C, his speed was zero. So Craig did not move at all. He may have stopped to eat.

Since Craig travels a total of 230 miles in 7 hours, his average speed is $\frac{230}{7} \approx 32.9$ miles per hour.

35. Start by setting $g(x)$ and $2k(x)$ equal to each other:

$$g(x) = 2k(x)$$
$$2x^2 + 3x + 10 = 2(2x + 16)$$
$$2x^2 + 3x + 10 = 4x + 32$$
$$-4x \qquad = -4x$$
$$2x^2 - x + 10 = 32$$
$$-32 = -32$$
$$2x^2 - x - 22 = 0$$

Then use the quadratic formula:

$$x = \frac{-b \pm \sqrt{b^2 - 4ac}}{2a}$$

$$x = \frac{-(-1) \pm \sqrt{(-1)^2 - 4 \cdot 2 \cdot (-22)}}{2 \cdot 2}$$

$$x = \frac{1 \pm \sqrt{1 + 176}}{4} = \frac{1 \pm \sqrt{177}}{4}$$

$$x = \frac{1 + \sqrt{177}}{4} \approx 3.6 \quad \text{or} \quad x = \frac{1 - \sqrt{177}}{4} \approx -3.1$$

Of the three options: factoring, completing the square, and the quadratic formula, the quadratic formula was simplest for this example. The coefficient of 2 in front of the x^2 makes using the other two options complicated.

36. A function to model Option 1 is $f(x) = 10 + 100x$.
    A function to model Option 2 is $g(x) = 10 \cdot 2^x$.

    After 7 weeks with Option 1, he will have:

    $$f(7) = 10 + 100 \cdot 7$$
    $$= 10 + 700$$
    $$= 710$$

    After 7 weeks with Option 2, he will have:

    $$g(7) = 10 \cdot 2^7$$
    $$= 10 \cdot 128$$
    $$= 1280$$

    Both Option 1 and Option 2 will enable Michael to have more than \$700. Either option will let him reach his goal.

PART IV

37. The equation for the swim team can be written as $y = 5 + 10x$ or $y = 10x + 5$.

    The equation for the chorus can be written as $y = 35 + 5x$ or $y = 5x + 35$.

    Before graphing, choose an appropriate scale. Since the point of intersection must be on the graph, you must find that point of intersection before deciding on the scale.

    Using a chart is the simplest way to determine the approximate intersection point.

x	swim	chorus
0	5	35
1	15	40
2	25	45
3	35	50
4	45	55
5	55	60
6	65	65
7	75	70

Since the grid is 20 by 20 and the point (6, 65) must be on it, one possible scale is to have one box equal 0.5 units on the x-axis and 5 units on the y-axis.

The graph looks like this:

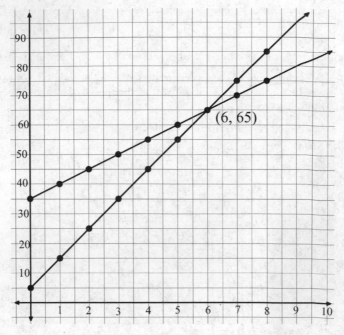

The point of intersection is (6, 65). It means that in the year 2016, both the swim team and the chorus had 65 members.

Topic	Question Numbers	Number of Points	Your Points	Your Percentage
1. Polynomials	5, 10, 25, 31, 33	2 + 2+ 2 + 2 + 4 = 12		
2. Properties of Algebra	2, 19, 23	2 + 2 + 2 = 6		
3. Functions	9, 17, 20, 21, 24, 32, 35	2 + 2+ 2 + 2 + 2 + 2 + 4 = 16		
4. Creating and Interpreting Equations	4, 28	2 + 2 = 4		
5. Inequalities	11, 13, 30	2 + 2 + 2 = 6		
6. Sequences and Series	18	2		
7. Systems of Equations	8, 37	2 + 6 = 8		
8. Quadratic Equations and Factoring	6, 22	2 + 2 = 4		
9. Regression	14	2		
10. Exponential Equations	7, 12, 36	2 + 2 + 4 = 8		
11. Graphing	1, 16, 26, 34	2 + 2 + 2 + 4 = 10		
12. Statistics	15, 29	2 + 2 = 4		
13. Number Properties	20, 27	2 + 2 = 4		

HOW TO CONVERT YOUR RAW SCORE TO YOUR ALGEBRA I REGENTS EXAMINATION SCORE

The accompanying conversion chart must be used to determine your final score on the June 2017 Regents Examination in Algebra I. To find your final exam score, locate in the column labeled "Raw Score" the total number of points you scored out of a possible 86 points. Since partial credit is allowed in Parts II, III, and IV of the test, you may need to approximate the credit you would receive for a solution that is not completely correct. Then locate in the adjacent column to the right the scale score that corresponds to your raw score. The scale score is your final Algebra I Regents Examination score.

Regents Examination in Algebra I—June 2017
Chart for Converting Total Test Raw Scores to Final
Examination Scores (Scaled Scores)

Raw Score	Scale Score	Performance Level	Raw Score	Scale Score	Performance Level	Raw Score	Scale Score	Performance Level
86	100	5	57	81	4	28	66	3
85	99	5	56	81	4	27	65	3
84	97	5	55	80	4	26	64	2
83	96	5	54	80	4	25	62	2
82	95	5	53	80	4	24	61	2
81	94	5	52	80	4	23	60	2
80	93	5	51	79	3	22	58	2
79	92	5	50	79	3	21	57	2
78	91	5	49	79	3	20	55	2
77	90	5	48	78	3	19	53	1
76	90	5	47	78	3	18	52	1
75	89	5	46	78	3	17	50	1
74	88	5	45	77	3	16	48	1
73	88	5	44	77	3	15	46	1
72	87	5	43	77	3	14	44	1
71	86	5	42	76	3	13	42	1
70	86	5	41	76	3	12	39	1
69	86	5	40	75	3	11	37	1
68	85	5	39	75	3	10	34	1
67	84	4	38	74	3	9	32	1
66	84	4	37	73	3	8	29	1
65	83	4	36	73	3	7	26	1
64	83	4	35	72	3	6	23	1
63	83	4	34	71	3	5	19	1
62	82	4	33	70	3	4	16	1
61	82	4	32	70	3	3	12	1
60	82	4	31	69	3	2	8	1
59	82	4	30	68	3	1	4	1
58	81	4	29	67	3	0	0	1

Examination August 2017
Algebra I

HIGH SCHOOL MATH REFERENCE SHEET

Conversions

1 inch = 2.54 centimeters	1 cup = 8 fluid ounces
1 meter = 39.37 inches	1 pint = 2 cups
1 mile = 5280 feet	1 quart = 2 pints
1 mile = 1760 yards	1 gallon = 4 quarts
1 mile = 1.609 kilometers	1 gallon = 3.785 liters
	1 liter = 0.264 gallon
1 kilometer = 0.62 mile	1 liter = 1000 cubic centimeters
1 pound = 16 ounces	
1 pound = 0.454 kilogram	
1 kilogram = 2.2 pounds	
1 ton = 2000 pounds	

Formulas

Triangle	$A = \frac{1}{2}bh$
Parallelogram	$A = bh$
Circle	$A = \pi r^2$
Circle	$C = \pi d$ or $C = 2\pi r$

Formulas (continued)

General Prisms	$V = Bh$
Cylinder	$V = \pi r^2 h$
Sphere	$V = \frac{4}{3}\pi r^3$
Cone	$V = \frac{1}{3}\pi r^2 h$
Pyramid	$V = \frac{1}{3}Bh$
Pythagorean Theorem	$a^2 + b^2 = c^2$
Quadratic Formula	$x = \dfrac{-b \pm \sqrt{b^2 - 4ac}}{2a}$
Arithmetic Sequence	$a_n = a_1 + (n-1)d$
Geometric Sequence	$a_n = a_1 r^{n-1}$
Geometric Series	$S_n = \dfrac{a_1 - a_1 r^n}{1-r}$ where $r \neq 1$
Radians	1 radian $= \dfrac{180}{\pi}$ degrees
Degrees	1 degree $= \dfrac{\pi}{180}$ radians
Exponential Growth/Decay	$A = A_0 e^{k(t-t_0)} + B_0$

PART I

Answer all 24 questions in this part. Each correct answer will receive 2 credits. No partial credit will be allowed. For each statement or question, write in the space provided the numeral preceding the word or expression that best completes the statement or answers the question.   [48 credits]

1 A part of Jennifer's work to solve the equation $2(6x^2 - 3) = 11x^2 - x$ is shown below.

Given: $2(6x^2 - 3) = 11x^2 - x$
Step 1: $12x^2 - 6 = 11x^2 - x$

Which property justifies her first step?

(1) identity property of multiplication
(2) multiplication property of equality
(3) commutative property of multiplication
(4) distributive property of multiplication over subtraction

1 ____

2 Which value of x results in equal outputs for $j(x) = 3x - 2$ and $b(x) = |x + 2|$?

(1) –2   (3) $\frac{2}{3}$

(2) 2   (4) 4

2 ____

3 The expression $49x^2 - 36$ is equivalent to

(1) $(7x - 6)^2$ (3) $(7x - 6)(7x + 6)$
(2) $(24.5x - 18)^2$ (4) $(24.5x - 18)(24.5x + 18)$

3 ____

4 If $f(x) = \frac{1}{2}x^2 - \left(\frac{1}{4}x + 3\right)$, what is the value of $f(8)$?

(1) 11   (3) 27
(2) 17   (4) 33

4 ____

5 The graph below models the height of a remote-control helicopter over 20 seconds during flight.

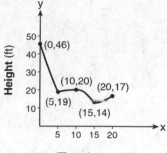

Time (seconds)

Over which interval does the helicopter have the *slowest* average rate of change?

(1) 0 to 5 seconds   (3) 10 to 15 seconds

(2) 5 to 10 seconds   (4) 15 to 20 seconds   5 _____

6 In the functions $f(x) = kx^2$ and $g(x) = |kx|$, k is a positive integer. If k is replaced by $\frac{1}{2}$, which statement about these new functions is true?

(1) The graphs of both $f(x)$ and $g(x)$ become wider.

(2) The graph of $f(x)$ becomes narrower and the graph of $g(x)$ shifts left.

(3) The graphs of both $f(x)$ and $g(x)$ shift vertically.

(4) The graph of $f(x)$ shifts left and the graph of $g(x)$ becomes wider.   6 _____

7 Wenona sketched the polynomial $P(x)$ as shown on the axes below.

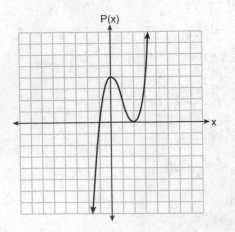

Which equation could represent $P(x)$?

(1) $P(x) = (x + 1)(x - 2)^2$
(2) $P(x) = (x - 1)(x + 2)^2$
(3) $P(x) = (x + 1)(x - 2)$
(4) $P(x) = (x - 1)(x + 2)$

7 _____

8 Which situation does *not* describe a causal relationship?

(1) The higher the volume on a radio, the louder the sound will be.
(2) The faster a student types a research paper, the more pages the research paper will have.
(3) The shorter the time a car remains running, the less gasoline it will use.
(4) The slower the pace of a runner, the longer it will take the runner to finish the race.

8 _____

9 A plumber has a set fee for a house call and charges by the hour for repairs. The total cost of her services can be modeled by $c(t) = 125t + 95$.

Which statements about this function are true?

I.   A house call fee costs $95.
II. The plumber charges $125 per hour.
III. The number of hours the job takes is represented by t.

(1) I and II, only   (3) II and III, only
(2) I and III, only   (4) I, II, and III   9 _____

10 What is the domain of the relation shown below?

$$\{(4, 2),(1, 1),(0, 0),(1, -1),(4, -2)\}$$

(1) $\{0, 1, 4\}$
(2) $\{-2, -1, 0, 1, 2\}$
(3) $\{-2, -1, 0, 1, 2, 4\}$
(4) $\{-2, -1, 0, 0, 1, 1, 1, 2, 4, 4\}$ 10 _____

11 What is the solution to the inequality $2 + \dfrac{4}{9}x \geq 4 + x$?

(1) $x \leq -\dfrac{18}{5}$ (3) $x \leq \dfrac{54}{5}$

(2) $x \geq -\dfrac{18}{5}$ (4) $x \geq \dfrac{54}{5}$ 11 _____

$$18 + 4x \geq 36 + 9x$$
$$-36 \quad -4x \quad -36 \quad -4x$$
$$\frac{-18}{5} \geq 5x$$

12 Konnor wants to burn 250 Calories while exercising for 45 minutes at the gym. On the treadmill, he can burn 6 Cal/min. On the stationary bike, he can burn 5 Cal/min.

If t represents the number of minutes on the treadmill and b represents the number of minutes on the stationary bike, which expression represents the number of Calories that Konnor can burn on the stationary bike?

(1) b (3) $45 - b$

(2) $5b$ (4) $250 - 5b$

12

13 Which value of x satisfies the equation $\dfrac{5}{6}\left(\dfrac{3}{8} - x\right) = 16$?

(1) -19.575 (3) -16.3125

(2) -18.825 (4) -15.6875

13 _____

14 If a population of 100 cells triples every hour, which function represents $p(t)$, the population after t hours?

(1) $p(t) = 3(100)^t$ (3) $p(t) = 3t + 100$

(2) $p(t) = 100(3)^t$ (4) $p(t) = 100t + 3$

14

15 A sequence of blocks is shown in the diagram below.

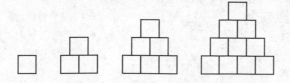

This sequence can be defined by the recursive function $a_1 = 1$ and $a_n = a_{n-1} + n$. Assuming the pattern continues, how many blocks will there be when $n = 7$?

(1) 13   (3) 28

(2) 21   (4) 36   15 _____

16 Mario's \$15,000 car depreciates in value at a rate of 19% per year. The value, V, after t years can be modeled by the function $V = 15{,}000(0.81)^t$. Which function is equivalent to the original function?

(1) $V = 15{,}000(0.9)^{9t}$ (3) $V = 15{,}000(0.9)^{\frac{t}{9}}$

(2) $V = 15{,}000(0.9)^{2t}$ (4) $V = 15{,}000(0.9)^{\frac{t}{2}}$ 16 _____

17 The highest possible grade for a book report is 100. The teacher deducts 10 points for each day the report is late.

Which kind of function describes this situation?

(1) linear   (3) exponential growth

(2) quadratic   (4) exponential decay   17 _____

18 The function $h(x)$, which is graphed below, and the function $g(x) = 2|x + 4| - 3$ are given.

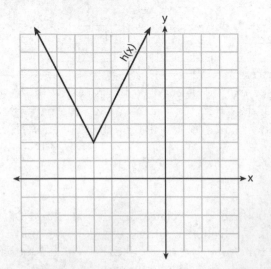

Which statements about these functions are true?

I. $g(x)$ has a lower minimum value than $h(x)$.
II. For all values of x, $h(x) < g(x)$.
III. For any value of x, $g(x) \neq h(x)$.

(1) I and II, only   (3) II and III, only
(2) I and III, only   (4) I, II, and III   18 _____

19 The zeros of the function $f(x) = 2x^3 + 12x - 10x^2$ are

(1) {2, 3}   (3) {0, 2, 3}
(2) {−1, 6}   (4) {0, −1, 6}   19 _____

20 How many of the equations listed below represent the line passing through the points $(2, 3)$ and $(4, -7)$?

$$5x + y = 13$$
$$y + 7 = -5(x - 4)$$
$$y = -5x + 13$$
$$y - 7 = 5(x - 4)$$

(1) 1   (3) 3

(2) 2   (4) 4   20 _____

21 The Ebola virus has an infection rate of 11% per day as compared to the SARS virus, which has a rate of 4% per day.

If there were one case of Ebola and 30 cases of SARS initially reported to authorities and cases are reported each day, which statement is true?

(1) At day 10 and day 53 there are more Ebola cases.
(2) At day 10 and day 53 there are more SARS cases.
(3) At day 10 there are more SARS cases, but at day 53 there are more Ebola cases.
(4) At day 10 there are more Ebola cases, but at day 53 there are more SARS cases.

       21 _____

22 The results of a linear regression are shown below.

$$y = ax + b$$
$$a = -1.15785$$
$$b = 139.3171772$$
$$r = -0.896557832$$
$$r^2 = 0.8038159461$$

Which phrase best describes the relationship between x and y?

(1) strong negative correlation
(2) strong positive correlation
(3) weak negative correlation
(4) weak positive correlation   22 _____

23 Abigail's and Gina's ages are consecutive integers. Abigail is younger than Gina and Gina's age is represented by x. If the difference of the square of Gina's age and eight times Abigail's age is 17, which equation could be used to find Gina's age?

(1) $(x + 1)^2 - 8x = 17$
(2) $(x - 1)^2 - 8x = 17$
(3) $x^2 - 8(x + 1) = 17$
(4) $x^2 - 8(x - 1) = 17$ 23 _____

24 Which system of equations does *not* have the same solution as the system below?

$$4x + 3y = 10$$
$$-6x - 5y = -16$$

(1) $-12x - 9y = -30$ (3) $24x + 18y = 60$
    $12x + 10y = 32$ $-24x - 20y = -64$

(2) $20x + 15y = 50$ (4) $40x + 30y = 100$
    $-18x - 15y = -48$ $36x + 30y = -96$ 24 _____

PART II

Answer all 8 questions in this part. Each correct answer will receive 2 credits. Clearly indicate the necessary steps, including appropriate formula substitutions, diagrams, graphs, charts, etc. For all questions in this part, a correct numerical answer with no work shown will receive only 1 credit. [16 credits]

25 A teacher wrote the following set of numbers on the board:

$$a = \sqrt{20} \qquad b = 2.5 \qquad c = \sqrt{225}$$

Explain why $a + b$ is irrational, but $b + c$ is rational.

26 Determine and state whether the sequence 1, 3, 9, 27, ... displays exponential behavior. Explain how you arrived at your decision.

Yes, not a constant rate of change.

27 Using the formula for the volume of a cone, express r in terms of V, h, and π.

$$3\left(V = \left(\frac{1}{3}\pi r^2 h\right)\right)3$$

$$\boxed{\sqrt{\frac{3V}{\pi h}}} = \pi r^2 h$$

28 The graph below models the cost of renting video games with a membership in Plan *A* and Plan *B*.

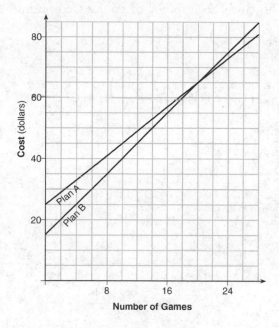

Explain why Plan *B* is the better choice for Dylan if he only has $50 to spend on video games, including a membership fee.

Bobby wants to spend $65 on video games, including a membership fee. Which plan should he choose? Explain your answer.

29 Samantha purchases a package of sugar cookies. The nutrition label states that each serving size of 3 cookies contains 160 Calories. Samantha creates the graph below showing the number of cookies eaten and the number of Calories consumed.

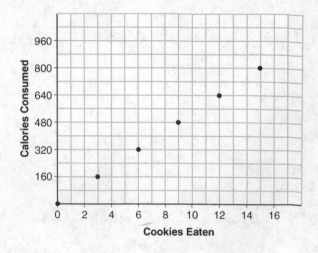

Explain why it is appropriate for Samantha to draw a line through the points on the graph.

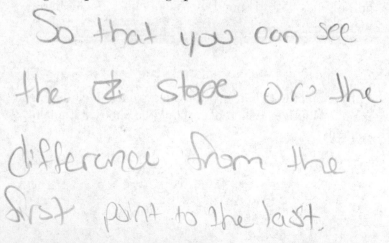

So that you can see the slope or the difference from the first point to the last.

30 A two-inch-long grasshopper can jump a horizontal distance of 40 inches. An athlete, who is five feet nine, wants to cover a distance of one mile by jumping. If this person could jump at the same ratio of body-length to jump-length as the grasshopper, determine, to the *nearest jump*, how many jumps it would take this athlete to jump one mile.

31 Write the expression $5x + 4x^2(2x + 7) - 6x^2 - 9x$ as a polynomial in standard form.

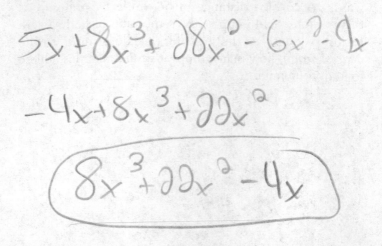

$5x + 8x^3 + 28x^2 - 6x^2 - 9x$

$-4x + 8x^3 + 22x^2$

$8x^3 + 22x^2 - 4x$

32 Solve the equation $x^2 - 6x = 15$ by completing the square.

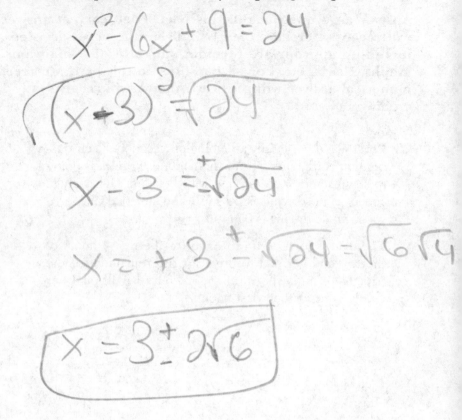

$$x^2 - 6x + 9 = 24$$

$$(x - 3)^2 = 24$$

$$x - 3 = \pm\sqrt{24}$$

$$x = +3 \pm \sqrt{24} = \sqrt{6}\sqrt{4}$$

$$x = 3 \pm 2\sqrt{6}$$

PART III

Answer all 4 questions in this part. Each correct answer will receive 4 credits. Clearly indicate the necessary steps, including appropriate formula substitutions, diagrams, graphs, charts, etc. For all questions in this part, a correct numerical answer with no work shown will receive only 1 credit. [16 credits]

33 Loretta and her family are going on vacation. Their destination is 610 miles from their home. Loretta is going to share some of the driving with her dad. Her average speed while driving is 55 mph and her dad's average speed while driving is 65 mph.

The plan is for Loretta to drive for the first 4 hours of the trip and her dad to drive for the remainder of the trip. Determine the number of hours it will take her family to reach their destination.

After Loretta has been driving for 2 hours, she gets tired and asks her dad to take over. Determine, to the *nearest tenth of an hour*, how much time the family will save by having Loretta's dad drive for the remainder of the trip.

34 The heights, in feet, of former New York Knicks basketball players are listed below.

6.4	6.9	6.3	6.2	6.3	6.0	6.1	6.3	6.8	6.2
6.5	7.1	6.4	6.3	6.5	6.5	6.4	7.0	6.4	6.3
6.2	6.3	7.0	6.4	6.5	6.5	6.5	6.0	6.2	

Using the heights given, complete the frequency table below.

Interval	Frequency
6.0 – 6.1	
6.2 – 6.3	
6.4 – 6.5	
6.6 – 6.7	
6.8 – 6.9	
7.0 – 7.1	

Question 34 is continued on the next page.

Question 34 continued.

Based on the frequency table created, draw and label a frequency histogram on the grid below.

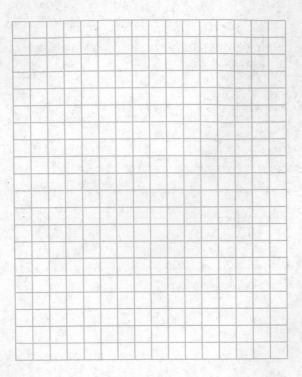

Determine and state which interval contains the upper quartile. Justify your response.

35 Solve the following system of inequalities graphically on
the grid below and label the solution *S*.

$$3x + 4y = 20$$
$$x < 3y - 18$$

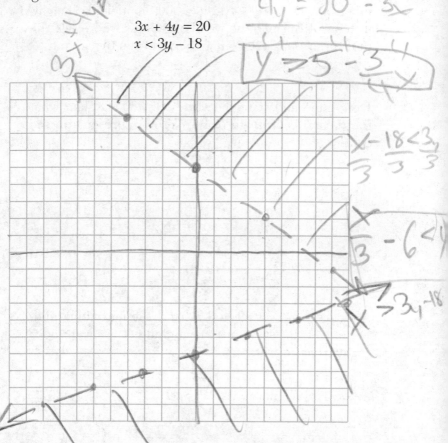

Is the point (3, 7) in the solution set? Explain your answer.

36 An Air Force pilot is flying at a cruising altitude of 9000 feet and is forced to eject from her aircraft. The function $h(t) = -16t^2 + 128t + 9000$ models the height, in feet, of the pilot above the ground, where t is the time, in seconds, after she is ejected from the aircraft.

Determine and state the vertex of $h(t)$. Explain what the second coordinate of the vertex represents in the context of the problem.

After the pilot was ejected, what is the maximum number of feet she was above the aircraft's cruising altitude? Justify your answer.

PART IV

Answer the question in this part. A correct answer will receive 6 credits. Clearly indicate the necessary steps, including appropriate formula substitutions, diagrams, graphs, charts, etc. A correct numerical answer with no work shown will receive only 1 credit. [6 credits]

37 Zeke and six of his friends are going to a baseball game. Their combined money totals $28.50. At the game, hot dogs cost $1.25 each, hamburgers cost $2.50 each, and sodas cost $0.50 each. Each person buys one soda. They spend all $28.50 on food and soda.

Write an equation that can determine the number of hot dogs, x, and hamburgers, y, Zeke and his friends can buy.

Question 37 is continued on the next page.

Question 37 continued.

Graph your equation on the grid below.

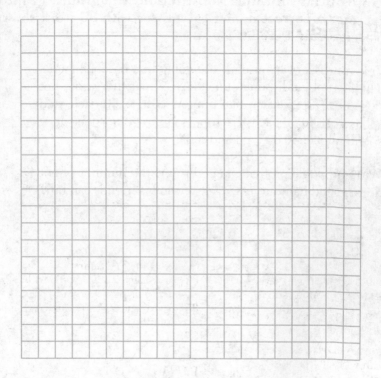

Determine how many different combinations, including those combinations containing zero, of hot dogs and hamburgers Zeke and his friends can buy, spending all $28.50. Explain your answer.

Answers
August 2017
Algebra I

Answer Key

PART I

1. (4)	**5.** (2)	**9.** (4)	**13.** (2)	**17.** (1)	**21.** (3)
2. (2)	**6.** (1)	**10.** (1)	**14.** (2)	**18.** (2)	**22.** (1)
3. (3)	**7.** (1)	**11.** (1)	**15.** (3)	**19.** (3)	**23.** (4)
4. (3)	**8.** (2)	**12.** (2)	**16.** (2)	**20.** (3)	**24.** (4)

PART II

25. Rational + Irrational = Irrational, Rational + Rational = Rational

26. Yes, the common ratio is 3.

27. $r = \sqrt{\dfrac{3V}{\pi h}}$

28. For \$50, Dylan can get 14 games with B compared to 12 games with A. For \$65, either plan gets Bobby 20 games.

29. She does not have to eat cookies 3 at a time. She can also eat a fraction of a cookie.

30. 46 jumps

31. $8x^3 + 22x^2 - 4x$

32. $x = 3 \pm \sqrt{24}$ or $x = 3 \pm 2\sqrt{6}$

PART III

33. 10 hours, 0.3 hours

34. See diagram, the top quartile is in the 6.4–6.5 interval

35. See diagram, (3, 7) is not in the solution set

36. (4, 9,256), 256 feet

PART IV

37. $1.25x + 2.50y = 25$, 11 different combinations

In Parts II–IV, you are required to show how you arrived at your answers. For sample methods of solutions, see the *Answers Explained* section.

Answers Explained

PART I

1. The distributive property of multiplication over subtraction says that an expression in the form $a \cdot (b - c)$ is equivalent to the expression $a \cdot b - a \cdot c$. In the solution, this property is applied to change $2(6x^2 - 3)$ to $2 \cdot 6x^2 - 2 \cdot 3 = 12x^2 - 6$.

   The correct choice is **(4)**.

2. Test each of the answer choices to see which one makes $j(x) = b(x)$.

   - Choice (1):

   $$j(-2) = 3 \cdot (-2) - 2$$
   $$= -6 - 2$$
   $$= -8$$

   $$b(-2) = |-2 + 2|$$
   $$= |0|$$
   $$= 0$$

   - Choice (2):

   $$j(2) = 3 \cdot 2 - 2$$
   $$= 6 - 2$$
   $$= 4$$

   $$b(2) = |2 + 2|$$
   $$= |4|$$
   $$= 4$$

   The correct choice is **(2)**.

3. The difference of perfect squares factoring pattern says that an expression of the form $a^2 - b^2$ is equivalent to the expression $(a - b)(a + b)$. The expression $49x^2 - 36$ can be rewritten as $(7x)^2 - 6^2$. Since this is a difference of perfect squares, it is equal to $(7x - 6)(7x + 6)$.

   The correct choice is **(3)**.

4. Replace each x with the number 8 and simplify:

$$f(8) = \frac{1}{2} \cdot 8^2 - \left(\frac{1}{4} \cdot 8 + 3\right)$$

$$= \frac{1}{2} \cdot 64 - (2 + 3)$$

$$= 32 - 5$$

$$= 27$$

The correct choice is **(3)**.

5. The average rate of change over an interval is the slope of the line segment that joins the endpoints of that interval. The slowest rate of change corresponds to the interval whose slope has the smallest absolute value. Even without doing any calculations, it can be seen that the interval from 5 to 10 seconds has the slope with the smallest absolute value since a line segment joining (5, 19) and (10, 20) is closest to being a horizontal line segment.

The average rate of change can also be calculated by using the slope formula $m = \frac{y_2 - y_1}{x_2 - x_1}$ for each of the intervals.

- Choice (1):

$$m = \frac{19 - 46}{5 - 0}$$

$$= \frac{-27}{5}$$

$$= -5\frac{2}{5}$$

The absolute value is $5\frac{2}{5}$.

- Choice (2):

$$m = \frac{20 - 19}{10 - 5}$$

$$= \frac{1}{5}$$

The absolute value is $\frac{1}{5}$.

- Choice (3):

$$m = \frac{14 - 20}{15 - 10}$$

$$= \frac{-6}{5}$$

$$= -1\frac{1}{5}$$

The absolute value is $1\frac{1}{5}$.

- Choice (4):

$$m = \frac{17 - 14}{20 - 15}$$

$$= \frac{3}{5}$$

The absolute value is $\frac{3}{5}$.

The correct choice is **(2)**.

6. If you choose an initial integer value for k, such as $k = 2$, you can use a graphing calculator to compare the graphs of $f(x) = 2x^2$ to $f(x) = \frac{1}{2}x^2$ and the graphs of $g(x) = |2x|$ to $g(x) = \left|\frac{1}{2}x\right|$.

For the TI-84:

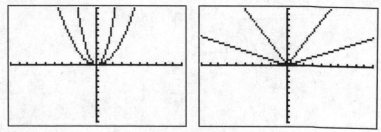

For the TI-Nspire:

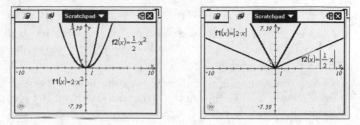

When k is replaced with $\frac{1}{2}$, each of the graphs becomes wider.

The correct choice is **(1)**.

7. Each of the answer choices can be graphed on the graphing calculator. When choice (1) is graphed, it looks like the graph from the question.

For the TI-84:

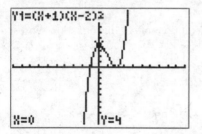

For the TI-Nspire:

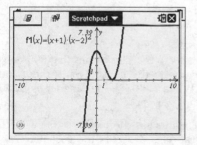

Even without the graphing calculator, it can be seen that the graph has an x-intercept at -1. So $(x + 1)$ would be one factor. The graph also has a double root at $x = 2$ since the graph bounces off the axis rather than passing through the axis. So there will be two $(x - 2)$ factors.

The correct choice is **(1)**.

8. A causal relationship is one where one event causes the other to happen.

   - Choice (1): Raising the volume does cause the radio to become louder.

   - Choice (2): How fast the student types does not matter. If the paper is five pages long, the student will complete typing the paper more quickly if he or she types faster, but the paper will not be longer.

   - Choice (3): Assuming that the car uses the same amount of gasoline each minute it runs, the shorter it remains running, the less gasoline the car will use.

   - Choice (4): Running slower does cause a runner to take more time to complete the race.

   There is an argument to be made that choice (2) could be considered causal if, for example, the student decides to spend exactly two hours on the paper and to type for that amount of time. In that case, the faster typist will have completed more pages after two hours. Compared to the other choices, though, choice (2) is still the best choice.

   The correct choice is **(2)**.

9. One way to determine what the components of this equation represent is to calculate the values of $c(0)$, $c(1)$, and $c(2)$:

   $$c(0) = 125 \cdot 0 + 95$$
   $$= 0 + 95$$
   $$= 95$$

   $$c(1) = 125 \cdot 1 + 95$$
   $$= 125 + 95$$
   $$= 220$$

   $$c(2) = 125 \cdot 2 + 95$$
   $$= 250 + 95$$
   $$= 345$$

   So when $t = 0$, the function evaluates to 95. This means that $95 must be the cost to get the plumber to come to the house even without spending any time doing the repairs. When $t = 1$, the fee increases by $125. So 125 must be the price per hour.

   In general, when a real-world scenario has the form $y = mx + b$, the b represents the fixed costs and the m represents the cost increase for every time x increases by 1.

   The correct choice is **(4)**.

10. The domain of a set of ordered pairs is a set containing each unique x-coordinate from each of the ordered pairs. The x-coordinates of the five ordered pairs are 4, 1, 0, 1, and 4. Since a set does not contain repeated numbers, this domain can be described as $\{0, 1, 4\}$.

The correct choice is **(1)**.

11. Start by solving the inequality the way you would solve an equality:

$$2 + \frac{4}{9}x \geq 4 + x$$

$$-x = -x$$

$$2 + \frac{4}{9}x - \frac{9}{9}x \geq 4$$

$$2 - \frac{5}{9}x \geq 4$$

$$-2 = -2$$

$$-\frac{5}{9}x \geq 2$$

For the last step, multiply both sides of the equation by $-\frac{9}{5}$. With inequalities, you have to switch the direction of the inequality sign anytime you multiply or divide both sides of the inequality by a negative:

$$\left(-\frac{9}{5}\right)\left(-\frac{5}{9}\right)x \leq \left(-\frac{9}{5}\right)2$$

$$x \leq -\frac{18}{5}$$

The correct choice is **(1)**.

12. If you were trying to solve for the number of minutes on each of the exercise machines, you would need to solve a system of equations:

$$t + b = 45$$

$$6t + 5b = 250$$

However, this question does not ask you to solve for t and b. It just asks how many calories Konnor burns on the bike. If he spends b hours on the bike and burns 5 Cal/min each minute on the bike, he burns $5b$ calories on the bike. The other information in the question is irrelevant.

The correct choice is **(2)**.

13. Solve the equation for x:

$$\frac{5}{6}\left(\frac{3}{8} - x\right) = 16$$

$$\left(\frac{5}{6}\right)\left(\frac{3}{8}\right) - \frac{5}{6}x = 16$$

$$\frac{15}{48} - \frac{5}{6}x = 16$$

$$-\frac{15}{48} = -\frac{15}{48}$$

$$-\frac{5}{6}x = 16 - \frac{15}{48}$$

$$= 15.6875$$

$$\left(-\frac{6}{5}\right)\left(-\frac{5}{6}x\right) = \left(-\frac{6}{5}\right) \cdot 15.6875$$

$$x = -18.825$$

The correct choice is (**2**).

14. The population of cells forms the sequence 100, 300, 900, 2,700, ...

If you substitute 0, 1, and 2 into the functions given in the choices, choice (2) becomes

$$p(0) = 100(3)^0$$
$$= 100 \cdot 1$$
$$= 100$$

$$p(1) = 100(3)^1$$
$$= 100 \cdot 3$$
$$= 300$$

$$p(2) = 100(3)^2$$
$$= 100 \cdot 9$$
$$= 900$$

An alternate way of answering this is to notice that since the numbers increase exponentially, the function must be of the form $p(t) = a \cdot b^t$. In a function of this form, a represents the initial value and b represents the growth factor. The initial value is 100, and the growth factor is 3. So the function is $p(t) = 100(3)^t$.

The correct choice is (**2**).

15. The given values of the sequence are $a_1 = 1$, $a_2 = 3$, $a_3 = 6$, and $a_4 = 10$. To get the next three values, substitute $n = 5$, 6, and 7 into the recursive function definition:

$$a_5 = a_{5-1} + 5$$
$$= a_4 + 5$$
$$= 10 + 5$$
$$= 15$$

$$a_6 = a_{6-1} + 6$$
$$= a_5 + 6$$
$$= 15 + 6$$
$$= 21$$

$$a_7 = a_{7-1} + 7$$
$$= a_6 + 7$$
$$= 21 + 7$$
$$= 28$$

The correct choice is **(3)**.

16. The simplest way to solve this question is to substitute $t = 1$ into each of the answer choices to see which gives the same answer as $V = 15,000(0.81)^1 = 15,000 \cdot 0.81 = 12,150$.

When you substitute $t = 1$ into choice (2), it becomes:

$$V = 15,000(0.9)^{2 \cdot 1}$$
$$= 15,000(0.9)^2$$
$$= 15,000 \cdot 0.81$$
$$= 12,150$$

Another way to see that choice (2) is equivalent to the original equation is to use the property of exponents that $x^{ab} = (x^a)^b$:

$$15,000 \cdot (0.9)^{2t} = 15,000(0.9^2)^t$$
$$= 15,000 \, (0.81)^t$$

The correct choice is **(2)**.

17. If you make a graph where for each point the x-coordinate is the number of days late and the y-coordinate is the highest possible grade, it will look like this:

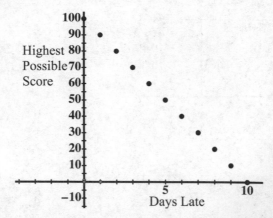

Since the points all fall on a line, this is a linear relationship.

The correct choice is **(1)**.

18. When you graph $g(x)$ on the same set of axes as $h(x)$, they look like this:

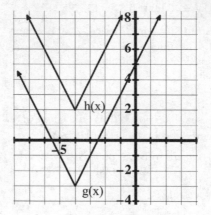

Using the graphing calculator, plot function g. It becomes the same as the graph of h but shifted down 5 units.

For the TI-84:

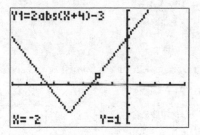

For the TI-Nspire:

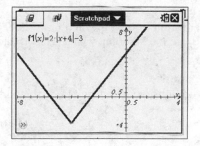

- Statement I: The minimum value of $g(x)$ is -3, while the minimum value of $h(x)$ is 2. So the minimum value of $g(x)$ is lower than the minimum value of $h(x)$.

- Statement II: For every x-value, the y-coordinate of the graph of $h(x)$ is above the y-coordinate of the graph of $g(x)$. So it is not true that for all (or any) values of x that $h(x) < g(x)$.

- Statement III: Since $g(x)$ is the same as $h(x)$ but just shifted down 5 units, there is no x-value where $g(x) = h(x)$. So it is true that for any value of x, $g(x) \neq h(x)$.

Statements I and III are both true.

The correct choice is (2).

19. The zeros of a function are the values that make the function evaluate to zero. To calculate the zeros of f, find all the solutions to the equation $0 = 2x^3 + 12x - 10x^2$:

$$0 = 2x^3 + 12x - 10x^2$$
$$0 = 2x(x^2 + 6 - 5x)$$
$$= 2x(x^2 - 5x + 6)$$
$$= 2x(x - 2)(x - 3)$$

For the product to be equal to zero, $2x$, $x - 2$, or $x - 3$ must equal zero:

$$2x = 0$$
$$\frac{2x}{2} = \frac{0}{2} \quad \text{or}$$
$$x = 0$$

$$x - 2 = 0$$
$$+2 = +2 \quad \text{or}$$
$$x = 2$$

$$x - 3 = 0$$
$$+3 = +3$$
$$x = 3$$

The solution set is $\{0, 2, 3\}$.

If you divided both sides of the equation by x, you would have "lost" the $x = 0$ solution and found only $\{2, 3\}$ as your solution, which is incorrect.

The correct choice is **(3)**.

20. To find the equation of a line through two given points, first find the slope of the line. In this case, the slope is

$$m = \frac{-7 - 3}{4 - 2}$$
$$= \frac{-10}{2}$$
$$= -5$$

To get the equation of the line in slope-intercept form, pick one of the two points. Insert its coordinates and the slope you just calculated into the equation $y = mx + b$. Then solve for b:

$$y = mx + b$$
$$3 = -5 \cdot 2 + b$$
$$3 = -10 + b$$
$$+10 = +10$$
$$13 = b$$

In slope-intercept form, the equation of the line is $y = -5x + 13$.

Of the four equations listed, this exactly matches the third. However, you must rearrange the others and then compare them to $y = -5x + 13$:

- First equation:

$$5x + y = 13$$
$$-5x = -5x$$
$$y = -5x + 13$$

✔

- Second equation:

$$y + 7 = -5(x - 4)$$
$$y + 7 = -5x + 20$$
$$-7 = -7$$
$$y = -5x + 13$$

✔

- Fourth equation:

$$y - 7 = 5(x - 4)$$
$$y - 7 = 5x - 20$$
$$+7 = +7$$
$$y = 5x - 13$$

NO

Of the four equations, the first three are equivalent to $y = -5x + 13$ while the fourth is not. Three of the four equations, then, represent the line passing through the two points.

The correct choice is **(3)**.

21. The formula for exponential growth is $y = a(1 + r)^x$, where a is the starting value and r is the growth rate.

For the Ebola virus the formula is

$$y = 1(1 + 0.11)^x$$
$$= 1(1.11)^x$$

For the SARS virus the formula is:

$$y = 30(1 + .04)^x$$
$$= 30(1.04)^x$$

At day 10, the number of Ebola cases is

$$y = 1(1.11)^{10}$$
$$= 2.84$$
$$\approx 3$$

At day 10, the number of SARS cases is

$$y = 30(1.04)^{10}$$
$$= 44.4$$
$$\approx 44$$

At day 53, the number of Ebola cases is

$$y = 1(1.11)^{53}$$
$$= 252.4$$
$$\approx 252$$

At day 53, the number of SARS cases is

$$y = 30(1.04)^{53}$$
$$= 239.8$$
$$\approx 240$$

At day 10, there are more SARS cases. At day 53, there are more Ebola cases.

The correct choice is (3).

22. This information tells you that the equation for the line of best fit for a scatter plot is $y = -1.15785x + 139.3171772$. Since the slope of the line is negative, the relationship between x and y is negative. So eliminate choices (3) and (4).

The strength of the correlation is based on the r^2-value. The closer that is to 1, the stronger the correlation. Since 0.8038159461 is pretty close to 1, this can be considered a strong negative correlation.

The correct choice is (1).

23. Gina's age is represented by x. If Abigail is younger than Gina and their ages are consecutive integers, Abigail must be one year younger than Gina. So Abigail's age can be represented by $x - 1$.

The square of Gina's age is x^2.
Eight times Abigail's age is $8(x - 1)$.

If the difference between these is 17, an equation that could be used to find Gina's age is $x^2 - 8(x - 1) = 17$.

The correct choice is (4).

24. In a system of equations, if you multiply both sides of one (or both) of the equations by a constant, the new system will have the same solution as the original system. Look at each of the four choices.

   • Choice (1): This is what you would get if you multiplied both sides of the top equation by −3 and both sides of the bottom equation by −2.

   • Choice (2): This is what you would get if you multiplied both sides of the top equation by 5 and both sides of the bottom equation by 3.

   • Choice (3): This is what you would get if you multiplied both sides of the top equation by 6 and both sides of the bottom equation by 4.

   • Choice (4): Although you could get this top equation by multiplying both sides of the original top equation by 10, there is no way to get the bottom equation by multiplying both sides of the original bottom equation by the same number. If you multiply the left side of the bottom equation by −6 and the right side by +6, the new system of equations will have a different solution than the original system.

   The correct choice is **(4)**.

PART II

25. a is irrational because 20 is not a perfect square. b is rational because it is a terminating decimal. Even though c is a square root, it is rational because 225 is a perfect square and $\sqrt{225} = 15$.

$a + b$ is irrational because an irrational plus a rational is always irrational. $b + c$ is rational because a rational plus a rational is always rational.

Note: Sometimes an irrational plus an irrational is irrational while other times an irrational plus an irrational is a rational number.

26. A sequence in which each term is equal to the previous term multiplied by some constant exhibits exponential behavior. In this case, the geometric sequence 1, 3, 9, 27, ... has an initial term of 1 and a constant ratio of 3.

27. The formula for the volume of a cone is given in the reference sheet: $V = \frac{1}{3}\pi r^2 h$. To express r in terms of the other variables, treat the V and h as numbers and "solve" for r:

$$V = \frac{1}{3}\pi r^2 h$$

$$3 \cdot V = 3 \cdot \frac{1}{3}\pi r^2 h$$

$$3V = \pi r^2 h$$

$$\frac{3V}{\pi h} = \frac{\pi r^2 h}{\pi h}$$

$$\frac{3V}{\pi h} = r^2$$

$$\sqrt{\frac{3V}{\pi h}} = \sqrt{r^2}$$

$$\sqrt{\frac{3V}{\pi h}} = r$$

You should not include a ± symbol in front of the radical because the radius of a cone must be a positive number.

28. If you draw a horizontal line at the $50 mark on the vertical axis, you will see that the line for Plan *A* contains a point at approximately (12.25, 50) while the line for Plan *B* contains a point at (14, 50). This means that with Plan *A*, you can get only 12 games for $50 while with Plan *B* you can get 14. Plan *B* is a better choice for Dylan since he can get two additional games.

Because both lines contain the point (20, 65), it doesn't matter which plan Bobby chooses. With either plan, he can get 20 games for $65.

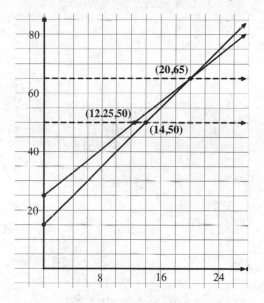

29. It is appropriate for Samantha to draw a line through the points on the graph because it is possible to eat a number of cookies that is not a multiple of 3. You can have 1 cookie or 2 cookies or even a half a cookie. The number of calories is proportional to the number of cookies. So one cookie has one-third the number of calories as 3 cookies, and half of a cookie has half as many calories as one cookie.

In mathematical terms, we say that the domain of the graph is all non-negative real numbers.

30. First set up a proportion to find the number of inches that the athlete can jump. Convert the athlete's height to inches so everything is expressed in the same unit:

$$5 \text{ feet } 9 \text{ inches} = (12 \times 5) + 9$$
$$= 60 + 9$$
$$= 69 \text{ inches}$$

$$\frac{2 \text{ in.}}{40 \text{ in.}} = \frac{69 \text{ in.}}{x \text{ in.}}$$
$$2x = 40 \cdot 69 = 2760$$
$$\frac{2x}{2} = \frac{2760}{2}$$
$$x = 1380$$

The athlete can jump 1,380 inches each jump.

To calculate the number of jumps the athlete needs to accomplish to make a mile, divide the number of inches in a mile by 1,380:

$$\text{Feet in a mile} = 5,280$$
$$\text{Inches in a mile} = 12 \times 5,280$$
$$= 63,360$$

$$63,360 \div 1,380 = 45.9$$
$$\approx 46 \text{ jumps}$$

31. First use the distributive property, and then combine the like terms:

$$5x + 4x^2(2x + 7) - 6x^2 - 9x$$
$$5x + 8x^3 + 28x^2 - 6x^2 - 9x$$
$$8x^3 + 28x^2 - 6x^2 + 5x - 9x$$
$$8x^3 + 22x^2 - 4x$$

32. To complete the square with an equation like $x^2 + bx = c$, add $\left(\dfrac{b}{2}\right)^2$ to both sides of the equation:

$$x^2 - 6x = 15$$
$$+\left(\frac{-6}{2}\right)^2 = +\left(\frac{-6}{2}\right)^2$$
$$x^2 - 6x + 9 = 15 + 9$$
$$x^2 - 6x + 9 = 24$$

The left side of this equation is now a perfect square and can be factored:

$$x^2 - 6x + 9 = 24$$
$$(x - 3)^2 = 24$$

Take the square root of both sides of the equation, and isolate the x-term:

$$(x - 3)^2 = 24$$
$$\sqrt{(x - 3)^2} = \pm\sqrt{24}$$
$$x - 3 = \pm\sqrt{24}$$
$$+3 = +3$$
$$x = 3 \pm \sqrt{24}$$

This can be simplified to $x = 3 \pm 2\sqrt{6}$.

PART III

33. In 4 hours, Loretta drives $4 \times 55 = 220$ miles. This leaves $610 - 220 = 390$ miles. At 65 mph, 390 miles can be traveled in $390 \div 65 = 6$ hours. This, together with Loretta's 4 hours of driving, is a total of 10 hours.

In 2 hours, Loretta drives $2 \times 55 = 110$ miles. This leaves $610 - 110 = 500$ miles. At 65 mph, 500 miles can be traveled in $500 \div 65 = 7.7$ hours. This, together with Loretta's 2 hours of driving, is a total of approximately 9.7 hours. This is 0.3 hours faster than the original plan.

34.

Interval	Frequency
6.0–6.1	3
6.2–6.3	10
6.4–6.5	11
6.6–6.7	0
6.8–6.9	2
7.0–7.1	3

A histogram is like a bar graph with no spaces between the bars. In a histogram you can choose the size of the intervals. Using an interval of 0.2, you get a histogram like this

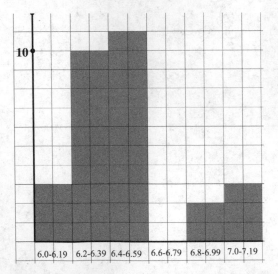

With 29 numbers, the 15th is the median. To get the top quartile, find the median of the top 14 numbers. Since the 7th and 8th largest numbers are both 6.5, the top quartile is 6.5, which is contained in the 6.4–6.5 interval.

35. Start by changing the inequalities into slope-intercept form.

First inequality:

$$3x + 4y > 20$$
$$-3x = -3x$$
$$4y > -3x + 20$$
$$\frac{4y}{4} > \frac{-3x + 20}{4}$$
$$y > -\frac{3}{4}x + 5$$

Second inequality:

$$x < 3y - 18$$
$$+18 = +18$$
$$x + 18 < 3y$$
$$\frac{x + 18}{3} < \frac{3y}{3}$$
$$\frac{1}{3}x + 6 < y$$
$$y > \frac{1}{3}x + 6$$

Graph the boundary line for the first inequality like $y = -\frac{3}{4}x + 5$ but as a dotted line because of the > sign. If there was a ≥ sign, the graph would need to be a solid line.

To decide which side of the line to shade, check if the ordered pair (0, 0) makes the inequality true:

$$0 > -\frac{3}{4} \cdot 0 + 5$$
$$0 > 0 + 5$$
$$0 > 5$$

NO

Since it is not true that $0 > 5$, shade the side of the line that does not contain $(0, 0)$.

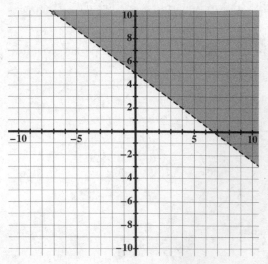

Graph the boundary line for the second inequality like $y = \dfrac{1}{3}x + 6$ but as a dotted line because of the $>$ sign.

To decide which side of the line to shade, check if the ordered pair $(0, 0)$ makes the inequality true:

$$0 > \frac{1}{3} \cdot 0 + 6$$
$$0 > 0 + 6$$
$$0 > 6$$

NO

Since it is not true that 0 > 6, shade the side of the line that does not contain (0, 0).

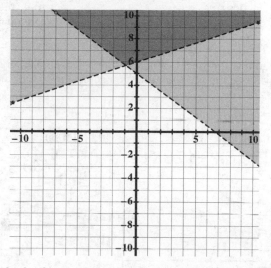

This can also be done with the graphing calculator.

For the TI-84:

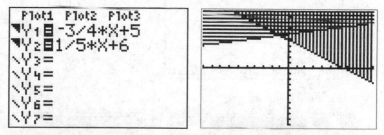

For the TI-Nspire:

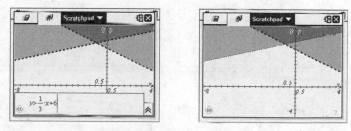

The point $(3, 7)$ is not in the solution set because it is on the dotted boundary line. You can also verify this by checking if $(3, 7)$ makes both inequalities true. Since $3 < 3 \cdot 7 - 18$ is not true (they are equal), the ordered pair $(3, 7)$ does not satisfy the second inequality. So $(3, 7)$ cannot be part of the solution set.

36. One way to find the vertex of $h(x)$ is to graph the function on the graphing calculator and to use the maximum command to find the coordinates of the vertex.

    For the TI-84:

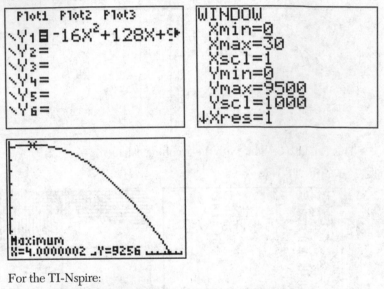

For the TI-Nspire:

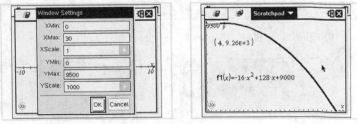

The x-coordinate of the vertex of a quadratic function (a parabola) can also

be found by using the formula $x = \dfrac{-b}{2a}$. For this function,

$x = \dfrac{-b}{2a} = \dfrac{-128}{2(-16)} = 4$. The y-coordinate of the vertex will be

$$h(4) = -16 \cdot 4^2 + 128 \cdot 4 + 9{,}000$$
$$= 9{,}256$$

The vertex of $h(t)$ is (4, 9,256).

Since cruising altitude is 9,000 feet and the maximum height achieved is 9,256 feet, the answer is 9,256 − 9,000 = 256 feet above the cruising altitude.

PART IV

37. Since there are 7 people and they each buy one soda, they spend $7 \times 0.50 = \$3.50$ on sodas. This leaves $\$28.50 - \$3.50 = \$25.00$ for the hot dogs and hamburgers.

    Since each hot dog costs \$1.25 and there are x hot dogs, the total cost of the hot dogs is $1.25x$. Since each hamburger costs \$2.50 and there are y hamburgers, the total cost of the hamburgers is $2.50y$. For the total cost of all the hot dogs and hamburgers to be \$25.00, the equation is

    $$1.25x + 2.50y = 25$$

    To graph this equation, convert to slope-intercept form:

    $$1.25x + 2.50y = 25$$
    $$-1.25x = -1.25x$$
    $$2.50y = -1.25x + 25$$
    $$\frac{2.50y}{2.50} = \frac{-1.25x + 25}{2.50}$$
    $$y = -\frac{1}{2}x + 10$$

    Graph this line on the grid.

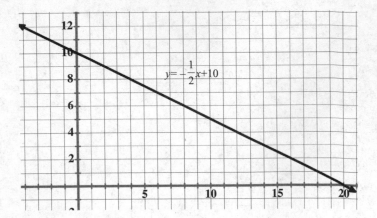

Since you cannot buy a fraction of a hot dog or a hamburger, find the coordinates of the points that have integers as both the x-coordinate and y-coordinate. There are 11 different combinations.

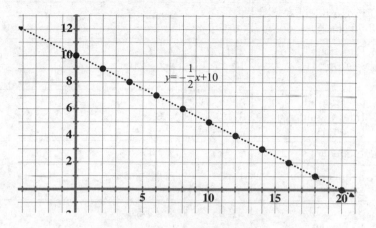

$$y=-\frac{1}{2}x+10$$

Hot Dogs	Hamburgers
20	0
18	1
16	2
14	3
12	4
10	5
8	6
6	7
4	8
2	9
0	10

Topic	Question Numbers	Number of Points	Your Points	Your Percentage
1. Polynomials	7, 31	2 + 2 = 4		
2. Properties of Algebra	1, 13, 27, 30	2 + 2 + 2 + 2 = 8		
3. Functions	2, 4, 5, 10, 17, 20, 19	2 + 2 + 2 + 2 + 2 + 2 + 2 = 14		
4. Creating and Interpreting Equations	9, 23, 33, 37	2 + 2 + 4 + 6 = 14		
5. Inequalities	11, 35	2 + 4 = 6		
6. Sequences and Series	15, 26	2 + 2 = 4		
7. Systems of Equations	24, 28	2 + 2 = 4		
8. Quadratic Equations and Factoring	3, 19, 32, 36	2 + 2 + 2 + 4 = 10		
9. Regression	22	2		
10. Exponential Equations	14, 16, 21	2 + 2 + 2 = 6		
11. Graphing	6, 18	2 + 2 = 4		
12. Statistics	8, 12, 24, 28, 34	2 + 2 + 2 + 2 + 4 = 12		
13. Number Properties	25	2		

HOW TO CONVERT YOUR RAW SCORE TO YOUR ALGEBRA I REGENTS EXAMINATION SCORE

The accompanying conversion chart must be used to determine your final score on the August 2017 Regents Examination in Algebra I. To find your final exam score, locate in the column labeled "Raw Score" the total number of points you scored out of a possible 86 points. Since partial credit is allowed in Parts II, III, and IV of the test, you may need to approximate the credit you would receive for a solution that is not completely correct. Then locate in the adjacent column to the right the scale score that corresponds to your raw score. The scale score is your final Algebra I Regents Examination score.

Regents Examination in Algebra I—August 2017
Chart for Converting Total Test Raw Scores to Final
Examination Scores (Scaled Scores)

Raw Score	Scale Score	Performance Level	Raw Score	Scale Score	Performance Level	Raw Score	Scale Score	Performance Level
86	100	5	57	81	4	28	66	3
85	98	5	56	81	4	27	65	3
84	97	5	55	81	4	26	64	2
83	96	5	54	80	4	25	63	2
82	94	5	53	80	4	24	62	2
81	93	5	52	80	4	23	60	2
80	92	5	51	80	3	22	59	2
79	91	5	50	79	3	21	57	2
78	91	5	49	79	3	20	56	2
77	90	5	48	79	3	19	55	1
76	89	5	47	78	3	18	52	1
75	88	5	46	78	3	17	50	1
74	88	5	45	78	3	16	49	1
73	87	5	44	77	3	15	47	1
72	87	5	43	77	3	14	44	1
71	86	5	42	77	3	13	42	1
70	86	5	41	76	3	12	40	1
69	86	5	40	76	3	11	37	1
68	85	5	39	75	3	10	35	1
67	84	4	38	74	3	9	32	1
66	84	4	37	74	3	8	29	1
65	84	4	36	73	3	7	26	1
64	83	4	35	73	3	6	23	1
63	83	4	34	72	3	5	19	1
62	83	4	33	71	3	4	16	1
61	82	4	32	70	3	3	12	1
60	82	4	31	69	3	2	8	1
59	82	4	30	68	3	1	4	1
58	82	4	29	67	3	0	0	1

Examination
June 2018
Algebra I

HIGH SCHOOL MATH REFERENCE SHEET

Conversions

1 inch = 2.54 centimeters

1 meter = 39.37 inches

1 mile = 5280 feet

1 mile = 1760 yards

1 mile = 1.609 kilometers

1 kilometer = 0.62 mile

1 pound = 16 ounces

1 pound = 0.454 kilogram

1 kilogram = 2.2 pounds

1 ton = 2000 pounds

1 cup = 8 fluid ounces

1 pint = 2 cups

1 quart = 2 pints

1 gallon = 4 quarts

1 gallon = 3.785 liters

1 liter = 0.264 gallon

1 liter = 1000 cubic centimeters

Formulas

Triangle	$A = \frac{1}{2}bh$
Parallelogram	$A = bh$
Circle	$A = \pi r^2$
Circle	$C = \pi d$ or $C = 2\pi r$

Formulas (continued)

General Prisms	$V = Bh$
Cylinder	$V = \pi r^2 h$
Sphere	$V = \dfrac{4}{3}\pi r^3$
Cone	$V = \dfrac{1}{3}\pi r^2 h$
Pyramid	$V = \dfrac{1}{3}Bh$
Pythagorean Theorem	$a^2 + b^2 = c^2$
Quadratic Formula	$x = \dfrac{-b \pm \sqrt{b^2 - 4ac}}{2a}$
Arithmetic Sequence	$a_n = a_1 + (n-1)d$
Geometric Sequence	$a_n = a_1 r^{n-1}$
Geometric Series	$S_n = \dfrac{a_1 - a_1 r^n}{1-r}$ where $r \neq 1$
Radians	$1 \text{ radian} = \dfrac{180}{\pi} \text{ degrees}$
Degrees	$1 \text{ degree} = \dfrac{\pi}{180} \text{ radians}$
Exponential Growth/Decay	$A = A_0 e^{k(t - t_0)} + B_0$

PART I

Answer all 24 questions in this part. Each correct answer will receive 2 credits. No partial credit will be allowed. For each statement or question, write in the space provided the numeral preceding the word or expression that best completes the statement or answers the question.   [48 credits]

1 The solution to $4p + 2 < 2(p + 5)$ is

  (1) $p > -6$ (3) $p > 4$

  (2) $p < -6$ (4) $p < 4$ 1 _____

2 If $k(x) = 2x^2 - 3\sqrt{x}$, then $k(9)$ is

  (1) 315   (3) 159

  (2) 307   (4) 153   2 _____

3 The expression $3(x^2 + 2x - 3) - 4(4x^2 - 7x + 5)$ is equivalent to

  (1) $-13x - 22x + 11$ (3) $19x^2 - 22x + 11$

  (2) $-13x^2 + 34x - 29$ (4) $19x^2 + 34x - 29$ 3 _____

4 The zeros of the function $p(x) = x^2 - 2x - 24$ are

  (1) -8 and 3   (3) -4 and 6

  (2) -6 and 4   (4) -3 and 8   4 _____

5 The box plot below summarizes the data for the average monthly high temperatures in degrees Fahrenheit for Orlando, Florida.

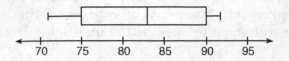

The third quartile is

(1) 92　　　　　　　(3) 83

(2) 90　　　　　　　(4) 71　　　　　　　5 _____

6 Joy wants to buy strawberries and raspberries to bring to a party. Strawberries cost \$1.60 per pound and raspberries cost \$1.75 per pound. If she only has \$10 to spend on berries, which inequality represents the situation where she buys x pounds of strawberries and y pounds of raspberries?

(1) $1.60x + 1.75y \le 10$　　(3) $1.75x + 1.60y \le 10$

(2) $1.60x + 1.75y \ge 10$　　(4) $1.75x + 1.60y \ge 10$　　6 _____

7 On the main floor of the Kodak Hall at the Eastman Theater, the number of seats per row increases at a constant rate. Steven counts 31 seats in row 3 and 37 seats in row 6. How many seats are there in row 20?

(1) 65　　　　　　　(3) 69

(2) 67　　　　　　　(4) 71　　　　　　　7 _____

8 Which ordered pair below is *not* a solution to
$f(x) = x^2 - 3x + 4$?

(1) $(0, 4)$ (3) $(5, 14)$

(2) $(1.5, 1.75)$ (4) $(-1, 6)$ 8 _____

9 Students were asked to name their favorite sport from a list of basketball, soccer, or tennis. The results are shown in the table below.

	Basketball	Soccer	Tennis
Girls	42	58	20
Boys	84	41	5

What percentage of the students chose soccer as their favorite sport?

(1) 39.6%   (3) 50.4%

(2) 41.4%   (4) 58.6%   9 _____

10 The trinomial $x^2 - 14x + 49$ can be expressed as

(1) $(x - 7)^2$ (3) $(x - 7)(x + 7)$

(2) $(x + 7)^2$ (4) $(x - 7)(x + 2)$ 10 _____

11 A function is defined as $\{(0, 1), (2, 3), (5, 8), (7, 2)\}$. Isaac is asked to create one more ordered pair for the function. Which ordered pair can he add to the set to keep it a function?

(1) $(0, 2)$ (3) $(7, 0)$

(2) $(5, 3)$ (4) $(1, 3)$ 11 _____

12 The quadratic equation $x^2 - 6x = 12$ is rewritten in the form $(x + p)^2 = q$, where q is a constant. What is the value of p?

(1) –12   (3) –3

(2) –9   (4) 9   12 _____

13 Which of the quadratic functions below has the *smallest* minimum value?

$$h(x) = x^2 + 2x - 6 \qquad\qquad k(x) = (x + 5)(x + 2)$$
          (1)   (3)

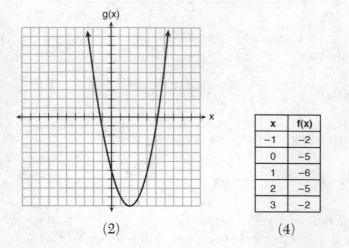

x	f(x)
–1	–2
0	–5
1	–6
2	–5
3	–2

          (2)   (4)   13 _____

14 Which situation is *not* a linear function?

(1) A gym charges a membership fee of $10.00 down and $10.00 per month.

(2) A cab company charges $2.50 initially and $3.00 per mile.

(3) A restaurant employee earns $12.50 per hour.

(4) A $12,000 car depreciates 15% per year.   14 _____

15 The Utica Boilermaker is a 15-kilometer road race. Sara is signed up to run this race and has done the following training runs:

> I.   10 miles
> II.   44,880 feet
> III.   15,560 yards

Which run(s) are at least 15 kilometers?

(1) I, only   (3) I and III

(2) II, only   (4) II and III   15 _____

16 If $f(x) = x^2 + 2$, which interval describes the range of this function?

(1) $(-\infty, \infty)$ (3) $[2, \infty)$

(2) $[0, \infty)$ (4) $(-\infty, 2]$ 16 _____

17 The amount Mike gets paid weekly can be represented by the expression $2.50a + 290$, where a is the number of cell phone accessories he sells that week. What is the constant term in this expression and what does it represent?

(1) $2.50a$, the amount he is guaranteed to be paid each week

(2) $2.50a$, the amount he earns when he sells a accessories

(3) 290, the amount he is guaranteed to be paid each week

(4) 290, the amount he earns when he sells a accessories   17 _____

18 A cubic function is graphed on the set of axes below.

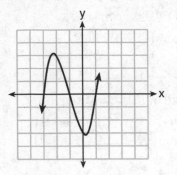

Which function could represent this graph?

(1) $f(x) = (x - 3)(x - 1)(x + 1)$
(2) $g(x) = (x + 3)(x + 1)(x - 1)$
(3) $h(x) = (x - 3)(x - 1)(x + 3)$
(4) $k(x) = (x + 3)(x + 1)(x - 3)$

18 _____

19 Mrs. Allard asked her students to identify which of the polynomials below are in standard form and explain why.

$$\text{I.} \quad 15x^4 - 6x + 3x^2 - 1$$
$$\text{II.} \quad 12x^3 + 8x + 4$$
$$\text{III.} \quad 2x^5 + 8x^2 + 10x$$

Which student's response is correct?

(1) Tyler said I and II because the coefficients are decreasing.
(2) Susan said only II because all the numbers are decreasing.
(3) Fred said II and III because the exponents are decreasing.
(4) Alyssa said II and III because they each have three terms.

19 _____

20 Which graph does *not* represent a function that is always increasing over the entire interval –2 < x < 2?

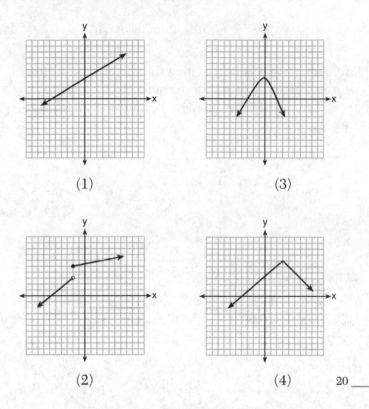

(1)

(2)

(3)

(4)   20 _____

21 At an ice cream shop, the profit, $P(c)$, is modeled by the function $P(c) = 0.87c$, where c represents the number of ice cream cones sold. An appropriate domain for this function is

(1) an integer ≤ 0   (3) a rational number ≤ 0
(2) an integer ≥ 0   (4) a rational number ≥ 0   21 _____

22 How many real-number solutions does $4x^2 + 2x + 5 = 0$ have?

(1) one   (3) zero

(2) two   (4) infinitely many   22 _____

23 Students were asked to write a formula for the length of a rectangle by using the formula for its perimeter, $p = 2\ell + 2w$. Three of their responses are shown below.

$$\text{I.} \quad \ell = \frac{1}{2}p - w$$

$$\text{II.} \quad \ell = \frac{1}{2}(p - 2w)$$

$$\text{III.} \quad \ell = \frac{p - 2w}{2}$$

Which responses are correct?

(1) I and II, only   (3) I and III, only

(2) II and III, only   (4) I, II, and III   23 _____

24 If $a_n = n(a_{n-1})$ and $a_1 = 1$, what is the value of a_5?

(1) 5   (3) 120

(2) 20   (4) 720   24 _____

PART II

Answer all 8 questions in this part. Each correct answer will receive 2 credits. Clearly indicate the necessary steps, including appropriate formula substitutions, diagrams, graphs, charts, etc. For all questions in this part, a correct numerical answer with no work shown will receive only 1 credit.   [16 credits]

25   Graph $f(x) = \sqrt{x+2}$ over the domain $-2 \le x \le 7$.

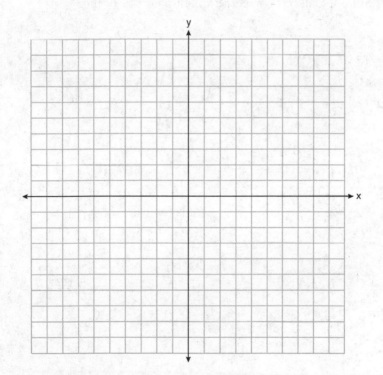

26 Caleb claims that the ordered pairs shown in the table below are from a nonlinear function.

x	f(x)
0	2
1	4
2	8
3	16

State if Caleb is correct. Explain your reasoning.

27 Solve for x to the *nearest tenth*: $x^2 + x - 5 = 0$.

28 The graph of the function $p(x)$ is represented below. On the same set of axes, sketch the function $p(x + 2)$.

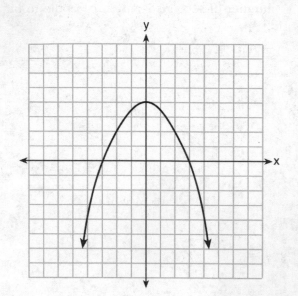

29 When an apple is dropped from a tower 256 feet high, the function $h(t) = -16t^2 + 256$ models the height of the apple, in feet, after t seconds. Determine, algebraically, the number of seconds it takes the apple to hit the ground.

30 Solve the equation below algebraically for the exact value of x.

$$6 - \frac{2}{3}(x + 5) = 4x$$

31 Is the product of $\sqrt{16}$ and $\frac{4}{7}$ rational or irrational?

Explain your reasoning.

32 On the set of axes below, graph the piecewise function:

$$f(x) = \begin{cases} -\dfrac{1}{2}x, & x < 2 \\[2mm] x, & x \geq 2 \end{cases}$$

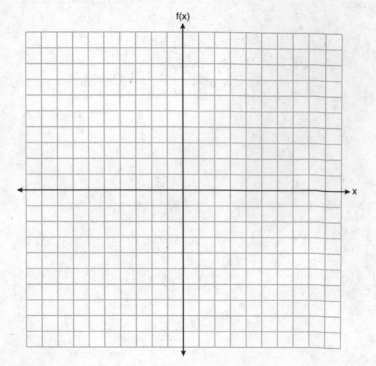

PART III

Answer all 4 questions in this part. Each correct answer will receive 4 credits. Clearly indicate the necessary steps, including appropriate formula substitutions, diagrams, graphs, charts, etc. For all questions in this part, a correct numerical answer with no work shown will receive only 1 credit. [16 credits]

33 A population of rabbits in a lab, $p(x)$, can be modeled by the function $p(x) = 20(1.014)^x$, where x represents the number of days since the population was first counted.

Explain what 20 and 1.014 represent in the context of the problem.

$20 \rightarrow$ Starting value of rabbits

$1.014 \rightarrow$ the growth factor

$1.4\% \rightarrow$ growth rate

Determine, to the *nearest tenth*, the average rate of change from day 50 to day 100.

$$\frac{Y_2 - Y_1}{X_2 - X_1} = \frac{80.32 - 40.08}{100 - 50} = \frac{40.24}{50}$$

$.8$

$(50, 4.08) \quad (100, 80.32)$

34 There are two parking garages in Beacon Falls. Garage A charges $7.00 to park for the first 2 hours, and each additional hour costs $3.00. Garage B charges $3.25 per hour to park.

When a person parks for at least 2 hours, write equations to model the cost of parking for a total of x hours in Garage A and Garage B.

Determine algebraically the number of hours when the cost of parking at both garages will be the same.

35 On the set of axes below, graph the following system of
inequalities:

$$2y + 3x \leq 14$$
$$4x - y < 2$$

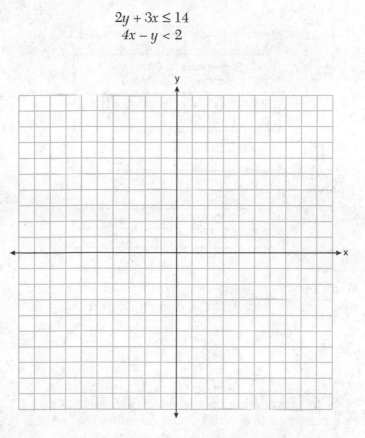

Determine if the point $(1, 2)$ is in the solution set.
Explain your answer.

36 The percentage of students scoring 85 or better on a mathematics final exam and an English final exam during a recent school year for seven schools is shown in the table below.

Percentage of Students Scoring 85 or Better	
Mathematics, x	English, y
27	46
12	28
13	45
10	34
30	56
45	67
20	42

Write the linear regression equation for these data, rounding all values to the *nearest hundredth*.

State the correlation coefficient of the linear regression equation, to the *nearest hundredth*. Explain the meaning of this value in the context of these data.

PART IV

Answer the question in this part. A correct answer will receive 6 credits. Clearly indicate the necessary steps, including appropriate formula substitutions, diagrams, graphs, charts, etc. A correct numerical answer with no work shown will receive only 1 credit. [6 credits]

37 Dylan has a bank that sorts coins as they are dropped into it. A panel on the front displays the total number of coins inside as well as the total value of these coins. The panel shows 90 coins with a value of $17.55 inside of the bank.

If Dylan only collects dimes and quarters, write a system of equations in two variables or an equation in one variable that could be used to model this situation.

Using your equation or system of equations, algebraically determine the number of quarters Dylan has in his bank.

Question 37 is continued on the next page.

Question 37 continued.

Dylan's mom told him that she would replace each one of his dimes with a quarter. If he uses all of his coins, determine if Dylan would then have enough money to buy a game priced at $20.98 if he must also pay an 8% sales tax. Justify your answer.

Answers
June 2018
Algebra I

Answer Key

PART I

1. (4)	**5.** (2)	**9.** (1)	**13.** (2)	**17.** (3)	**21.** (2)
2. (4)	**6.** (1)	**10.** (1)	**14.** (4)	**18.** (2)	**22.** (3)
3. (2)	**7.** (1)	**11.** (4)	**15.** (1)	**19.** (3)	**23.** (4)
4. (3)	**8.** (4)	**12.** (3)	**16.** (3)	**20.** (3)	**24.** (3)

PART II

25.

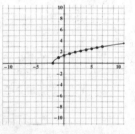

26. Caleb is correct. It is a nonlinear function.

27. $x = 1.8$ and $x = -2.8$

28. **See graph on page 504.**

29. 4 seconds

30. $\dfrac{4}{7}$

31. The product is rational.

32. **See graphs on pages 506 and 507.**

PART III

33. 20 is the starting population. 1.014 is the growth factor. The average rate of change is 0.8.

34. For Garage A, $A = 7 + 3(x - 2)$. For Garage B, $B = 3.25x$. Both garages cost the same when you park for 4 hours.

35. $(1, 2)$ is not in the solution set.

36. $y = 0.96x + 23.95$ The correlation coefficient is 0.92, which is a strong positive correlation.

PART IV

37. $0.10d + 0.25q = 17.55$

$d + q = 90$

Dylan has 57 quarters. No, he would not be able to buy the game.

In Parts II–IV, you are required to show how you arrived at your answers. For sample methods of solutions, see the *Answers Explained* **section.**

Answers Explained

PART I

1. Solving an inequality is very similar to solving an equality. Use the properties of algebra to isolate the variable.

   First distribute the 2 on the right-hand side:

   $$4p + 2 < 2(p + 5)$$
   $$4p + 2 < 2p + 10$$

   Then eliminate the constant term from the left-hand side:

   $$4p + 2 < 2p + 10$$
   $$-2 = -2$$
   $$4p < 2p + 8$$

   Next eliminate the variables from the right-hand side:

   $$4p < 2p + 8$$
   $$-2p = -2p$$
   $$2p < 8$$

   Finally, eliminate the coefficient by dividing both sides of the inequality by 2. If the coefficient is negative, the direction of the inequality would need to be reversed. In this case, the coefficient is positive. So do not reverse the inequality:

   $$\frac{2p}{2} < \frac{8}{2}$$
   $$p < 4$$

   The correct choice is **(4)**.

2. Substitute 9 for all occurrences of x in the function:

   $$k(9) = 2(9)^2 - 3\sqrt{9}$$
   $$= 2(81) - 3(3)$$
   $$= 162 - 9$$
   $$= 153$$

   The correct choice is **(4)**.

3. Distribute the 3 to the terms in the first parentheses:

$$3x^2 + 6x - 9 - 4(4x^2 - 7x + 5)$$

Then distribute the –4 to the terms in the second parentheses. Be sure to change the signs since you are multiplying by a negative:

$$3x^2 + 6x - 9 - 16x^2 + 28x - 20$$

Now combine like terms:

$$3x^2 - 16x^2 + 6x + 28x - 9 - 20 = -13x^2 + 34x - 29$$

The correct choice is **(2)**.

4. The zeros of a function are the numbers that evaluate to zero when you plug them into the function. To find the zeros, solve the equation $x^2 - 2x - 24 = 0$. There are several ways to solve this equation. The simplest is by factoring:

$$x^2 - 2x - 24 = 0$$
$$(x - 6)(x + 4) = 0$$
$$x - 6 = 0 \text{ or } x + 4 = 0$$
$$x = 6 \text{ or } x = -4$$

Another way to find the zeros of a function is to graph it on a graphing calculator and find the x-intercepts of the graph. This can be done with the zero function on the calculator.

For the TI-84:

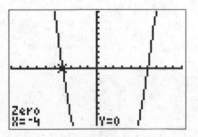

For the TI-Nspire:

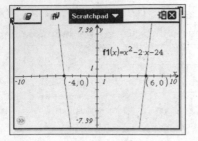

The correct choice is **(3)**.

5. In a box plot, the five values marked with vertical lines are the minimum, the first quartile, the median, the third quartile, and the maximum. For this box plot, the minimum is about 71, the first quartile is 75, the median is about 83, the third quartile is 90, and the maximum is about 92.

   The correct choice is **(2)**.

6. Since 1 pound of strawberries costs \$1.60, x pounds of strawberries cost $1.60x$. Since 1 pound of raspberries costs \$1.75, y pounds of raspberries cost $1.75y$. Together, the cost of x pounds of strawberries and y pounds of raspberries is $1.60x + 1.75y$. Since Joy has only \$10 to spend, the total cost must be less than or equal to 10. So the inequality that represents the situation is $1.60x + 1.75y \leq 10$.

   The correct choice is **(1)**.

7. One way to solve this is to make a list with the given information:

   Row 1.

   Row 2.

   Row 3. 31 seats

   Row 4.

   Row 5.

   Row 6. 37 seats

   The seats increase at a constant rate. Since the number of seats increases by 6 when the number of rows increases by 3, the increase in seats for each new row must be $6 \div 3 = 2$. Create a table to show the number of seats in each row:

1. 27	5. 35	9. 43	13. 51	17. 59
2. 29	6. 37	10. 45	14. 53	18. 61
3. 31	7. 39	11. 47	15. 55	19. 63
4. 33	8. 41	12. 49	16. 57	20. 65

A quicker way to solve this problem is to use the arithmetic sequence formula from the reference sheet, $a_n = a_1 + (n-1)d$. Remember that a_1 is the first term of the sequence, n is the term you want to find, and d is the difference from one term to the next.

For this question, $a_1 = 27$, $d = 2$, and $n = 20$. Plug the values into the formula:

$$\begin{aligned}
a_{20} &= 27 + (20 - 1)2 \\
&= 27 + 19 \cdot 2 \\
&= 27 + 38 \\
&= 65
\end{aligned}$$

A third way to solve this problem is to treat x as the row number and y as the number of seats in row x. Then find the equation of the line in slope-intercept form that passes through $(3, 31)$ and $(6, 37)$:

$$m = \frac{37 - 31}{6 - 3}$$

$$= \frac{6}{3}$$

$$= 2$$

$$y = mx + b$$

$$37 = 2 \cdot 6 + b$$

$$37 = 12 + b$$

$$-12 = -12$$

$$25 = b$$

$$y = 2x + 25$$

Substitute 20 for x:

$$\begin{aligned}
y &= 2 \cdot 20 + 25 \\
&= 40 + 25 \\
&= 65
\end{aligned}$$

The correct choice is **(1)**.

8. To check if an ordered pair is a solution, plug in the x-coordinate of the ordered pair and see if the function evaluates to the corresponding y-coordinate.

   • Testing choice (1):

$$f(0) = 0^2 - 3 \cdot 0 + 4 = 4 \checkmark$$

   • Testing choice (2):

$$f(1.5) = 1.5^2 - 3 \cdot 1.5 + 4 = 1.75 \checkmark$$

   • Testing choice (3):

$$f(5) = 5^2 - 3 \cdot 5 + 4 = 14 \checkmark$$

   • Testing choice (4):

$$f(-1) = (-1)^2 - 3 \cdot (-1) + 4 = 1 + 3 + 4 = 8 \neq 6$$

   The correct choice is **(4)**.

9. To calculate the percentage of students who chose soccer as their favorite sport, divide the number of students who chose soccer by the total number of students. There were 58 boys who chose soccer and 41 girls who chose soccer. So $58 + 41 = 99$ students chose soccer. The total number of students who answered the question was

$$42 + 58 + 20 + 84 + 41 + 5 = 250.$$

   The percentage of students who chose soccer was $\dfrac{99}{250} \approx 39.6\%$

   The correct choice is **(1)**.

10. You can factor an expression of the form $x^2 + bx + c$ into the product $(x + r_1)(x + r_2)$ by finding two numbers whose sum is b and whose product is c. In this problem, the two numbers whose sum is -14 and whose product is 49 are -7 and -7. So the expression factors into $(x - 7)(x - 7) = (x - 7)^2$.

Another way to solve this is to recognize that in an expression like $x^2 + bx + c$ when the c-term is equal to the square of half the b-term, the expression is a perfect square trinomial. That expression can be factored into $\left(x + \dfrac{b}{2}\right)^2$. In this example, $\left(\dfrac{-14}{2}\right)^2 = 49$. So the expression factors into $\left(x + \dfrac{-14}{2}\right)^2 = (x - 7)^2$.

Since this is a multiple-choice question, a third way this can be solved is to test each of the four choices to see which one becomes $x^2 - 14x + 49$ when multiplied.

- Testing choice (1):

$$\begin{aligned}(x - 7)^2 &= (x - 7)(x - 7) \\ &= x^2 - 7x - 7x + 49 \\ &= x^2 - 14x + 49\end{aligned}$$

- Testing choice (2):

$$\begin{aligned}(x + 7)^2 &= (x + 7)(x + 7) \\ &= x^2 + 7x + 7x + 49 \\ &= x^2 + 14x + 49\end{aligned}$$

- Testing choice (3):

$$\begin{aligned}(x - 7)(x + 7) &= x^2 + 7x - 7x - 49 \\ &= x^2 - 49\end{aligned}$$

- Testing choice (4):

$$\begin{aligned}(x - 7)(x + 2) &= x^2 + 2x - 7x - 14 \\ &= x^2 - 5x - 14\end{aligned}$$

The correct choice is **(1)**.

11. When a function is described as a set of ordered pairs, none of the x-coordinates can repeat.

- Testing choice (1): (0, 2) cannot be part of the function since the ordered pair (0, 1) already has 0 as an x-coordinate.

- Testing choice (2): (5, 3) cannot be part of the function since the ordered pair (5, 8) already has 5 as an x-coordinate.

- Testing choice (3): (7, 0) cannot be part of the function since the ordered pair (7, 2) already has 7 as an x-coordinate.

- Testing choice (4): (1, 3) can be part of the function since no other ordered pairs in the set have an x-coordinate of 1. It does not matter that the ordered pair (2, 3) already has a y-coordinate of 3.

The correct choice is **(4)**.

12. This question requires that you do the first two steps of solving a quadratic equation by completing the square process.

Step 1: The first step of solving an equation of the form $x^2 + bx = c$ by completing the square is to add $\left(\dfrac{b}{2}\right)^2$ to both sides of the equation.

For this example:

$$x^2 - 6x + \left(\dfrac{-6}{2}\right)^2 = 12 + \left(\dfrac{-6}{2}\right)^2$$
$$x^2 - 6x + 9 = 12 + 9$$
$$x^2 - 6x + 9 = 21$$

Step 2: Factor the left side of the equation:

$$(x - 3)(x - 3) = 21$$
$$(x - 3)^2 = 21$$

The completing the square process is not finished. Since this question just asks for the value of p, though, there is no more work to do. If $(x + p) = (x - 3)$, then p must be –3.

The correct choice is **(3)**.

13. The quadratic function with the smallest minimum value is the one whose vertex has the smallest y-coordinate.

- Testing choice (1): Use the graphing calculator to find the vertex of the parabola. The minimum is at $(-1, -7)$.

For the TI-84:

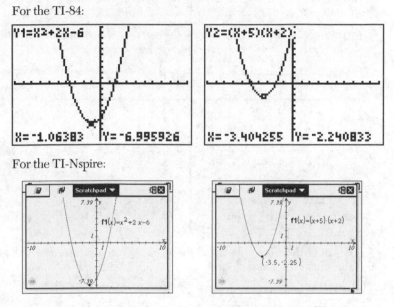

For the TI-Nspire:

- Testing choice (2): As shown on the graph, the vertex of the parabola is at $(2, -10)$.

- Testing choice (3): Use the graphing calculator to find the vertex of the parabola. The minimum is at $(-3.5, -2.25)$.

- Testing choice (4): When you graph the five given points, they resemble a parabola with a vertex at (1, –6).

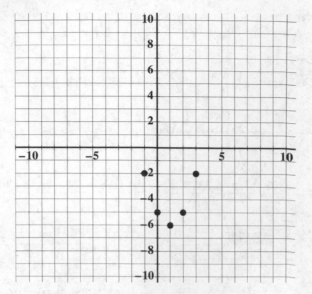

Since –10 is the minimum of all the *y*-coordinates of the four vertices, it is the smallest minimum value.

The correct choice is **(2)**.

14. One way to answer this question is to create an equation to model each scenario and determine which equation is not in the form $y = mx + b$.

- Testing choice (1): The equation is $y = 10x + 10$.

- Testing choice (2): The equation is $y = 3x + 2.5$.

- Testing choice (3): The equation is $y = 12.5x + 0$.

- Testing choice (4): The equation is $y = 12{,}000 \cdot 0.85^x$.

Choices (1), (2), and (3) are linear functions, while choice (4) is an exponential function.

Another way to answer this question is to make a sketch of each scenario and see which of them does not form a line.

- Testing choice (1):

- Testing choice (2):

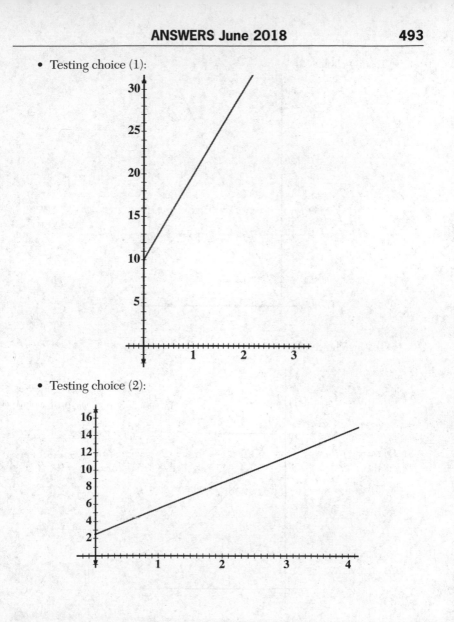

- Testing choice (3):

- Testing choice (4):

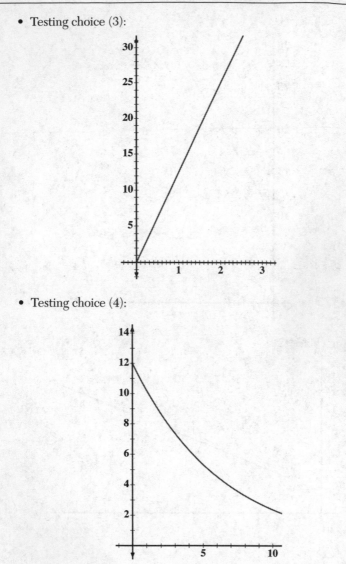

Only choice (4) does not form a line.

The correct choice is **(4)**.

15. Use the reference sheet to convert the measurements in each of the three runs into kilometers.

• Run I:

$$1 \text{ mile} = 1.609 \text{ kilometers}$$

$$10 \text{ miles}\left(\frac{1.609 \text{ kilometers}}{1 \text{ mile}}\right) = 16.09 \text{ kilometers}$$

$$16.09 \text{ kilometers} \geq 15 \text{ kilometers}$$

• Run II:

$$5280 \text{ feet} = 1 \text{ mile}$$

$$44{,}880 \text{ feet}\left(\frac{1 \text{ mile}}{5280 \text{ feet}}\right) = 8.5 \text{ miles}$$

$$1 \text{ mile} = 1.609 \text{ kilometers}$$

$$8.5 \text{ miles}\left(\frac{1.609 \text{ kilometers}}{1 \text{ mile}}\right) = 13.6765 \text{ kilometers} \approx 13.7 \text{ kilometers}$$

$$13.7 \text{ kilometers} < 15 \text{ kilometers}$$

• Run III:

$$1760 \text{ yards} = 1 \text{ mile}$$

$$15{,}560 \text{ yards}\left(\frac{1 \text{ mile}}{1760 \text{ yards}}\right) = 8.8409 \text{ miles} \approx 8.84 \text{ miles}$$

$$1 \text{ mile} = 1.609 \text{ kilometers}$$

$$8.84 \text{ miles}\left(\frac{1.609 \text{ kilometers}}{1 \text{ mile}}\right) = 14.22356 \text{ kilometers} \approx 14.2 \text{ kilometers}$$

$$14.2 \text{ kilometers} < 15 \text{ kilometers}$$

The correct choice is **(1)**.

16. The range of a function is the set of all the different y-coordinates of the graph of that function. You can graph this function on the graphing calculator.

For the TI-84:

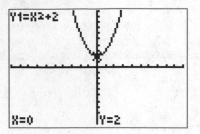

For the TI-Nspire:

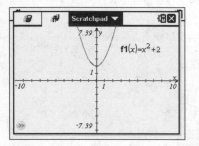

The minimum point has a y-coordinate of 2, and there is no maximum point. So the range is $y \geq 2$. In interval notation, this is written as $[2, \infty)$. The square bracket indicates that the interval includes the endpoint of 2. For the ∞, you always use a rounded parenthesis.

The correct choice is **(3)**.

17. In an expression in the form $mx + b$, the b is the constant term, the m is the coefficient, the x is the variable, and the mx is the variable part. In the expression $2.50a + 290$, the number 290 is the constant term.

When an expression like this is used to model a real-world scenario, you can interpret the meaning of the coefficient and of the constant by plugging in 0 and 1 for the variable.

If you plug in 0 for a, you get $2.50 \cdot 0 + 290 = 290$. So 290 is the amount Mike is guaranteed to be paid even if he sells 0 cell phone accessories.

If you plug in 1 for a, you get $2.50 \cdot 1 + 290 = 292.50$. So 2.50 must be the amount Mike earns for each accessory he sells. Therefore 2.50a is the amount Mike earns when he sells a accessories.

The correct choice is **(3)**.

18. If the graph of a polynomial function has an x-intercept at $(a, 0)$, the function will have a factor of $(x - a)$. Since the graph of this function has x-intercepts at $(-3, 0)$, $(-1, 0)$, and $(1, 0)$, the function must have factors of $(x -(-3)) = (x + 3)$, $(x -(-1)) = (x + 1)$, and $(x -(+1)) = (x - 1)$.

Another way to answer this is to graph the four choices to see which most resembles the given graph. On the graphing calculator, choice (2) looks most like the given graph.

For the TI-84:

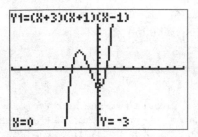

For the TI-Nspire:

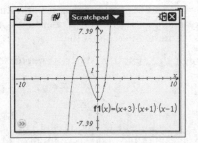

The correct choice is **(2)**.

19. A polynomial is in standard form if the exponents are decreasing from left to right. Whether the coefficients are decreasing, increasing, or fluctuating up and down makes no difference.

   - For option I, the exponents go from 4 to 1 to 2 to 0, which is not always decreasing.

   - For option II, the exponents decrease from 3 to 1 to 0.

   - For option III, the exponents decrease from 5 to 2 to 1.

   Both options II and III are in standard form because their exponents decrease in value. Both choices (3) and (4) identify which of the Roman numerals are in standard form. However, choice (4) is incorrect since the number of terms has nothing to do with whether or not a polynomial is in standard form.

   The correct choice is **(3)**.

20. A function is increasing over an interval if the graph is always going "up" as you move from left to right.

   - Testing choice (1): Between $x = -2$ and $x = 2$, the function is always increasing.

   - Testing choice (2): Between $x = -2$ and $x = 2$, the function is always increasing. There is a "jump" at $x = -2$. So it isn't accurate to say that this function is always increasing on all intervals. However between $x = -2$ and $x = 2$, it is increasing.

   - Testing choice (3): Between $x = -2$ and approximately $x = 0$, the function is increasing. From approximately $x = 0$ to $x = 2$, the function is decreasing.

   - Testing choice (4): Even though this function starts to decrease at $x = 3$, it is increasing from $x = -2$ to $x = 2$.

   The correct choice is **(3)**.

21. Since you cannot generally buy a fraction of an ice cream cone, the domain must be integers. Since you cannot buy a negative number of ice cream cones, the domain must be a nonnegative integer (including 0). If you were permitted to buy fractions of an ice cream cone, the answer would be choice (4).

   The correct choice is **(2)**.

22. The quickest way to solve this question is to graph the parabola $y = 4x^2 + 2x + 5$ and count how many x-intercepts it has.

For the TI-84:

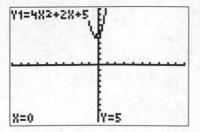

For the TI-Nspire:

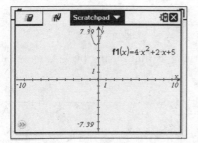

If you want to use algebra instead, calculate $b^2 - 4ac$. If $b^2 - 4ac = 0$, the equation has one real solution. If $b^2 - 4ac > 0$, the equation has two real solutions. If $b^2 - 4ac < 0$, the equation has no real solutions.

In this example, $b^2 - 4ac = 2^2 - 4 \cdot 4 \cdot 5 = 4 - 80 = -76 < 0$. So the equation has no real solutions.

The correct choice is **(3)**.

23. Isolate the l in the equation.

Step 1: Subtract $2w$ from both sides of the equation:

$$p = 2l + 2w$$
$$-2w = -2w$$
$$p - 2w = 2l$$

Step 2: Divide both sides by 2:

$$\frac{p - 2w}{2} = \frac{2l}{2}$$
$$\frac{p - 2w}{2} = l$$
$$l = \frac{p - 2w}{2}$$

This is option III.

Since dividing by 2 is the same thing as multiplying by $\frac{1}{2}$, the equation can also be written as:

$$\frac{1}{2}(p - 2w) = l$$
$$l = \frac{1}{2}(p - 2w)$$

This is option II.

If you distribute the $\frac{1}{2}$ through the right side of this equation you get:

$$l = \frac{1}{2}p - \frac{1}{2} \cdot 2w$$
$$l = \frac{1}{2}p - w$$

This is option I.

The correct choice is **(4)**.

24. To calculate a_5 with this recursive definition of the function, you first have to calculate a_2, a_3, and a_4.

For $n = 2$:

$$a_2 = 2(a_{2-1})$$
$$= 2(a_1)$$
$$= 2 \cdot 1$$
$$= 2$$

For $n = 3$:

$$a_3 = 3(a_{3-1})$$
$$= 3(a_2)$$
$$= 3 \cdot 2$$
$$= 6$$

For $n = 4$:

$$a_4 = 4(a_{4-1})$$
$$= 4(a_3)$$
$$= 4 \cdot 6$$
$$= 24$$

For $n = 5$:

$$a_5 = 5(a_{5-1})$$
$$= 5(a_4)$$
$$= 5 \cdot 24$$
$$= 120$$

The correct choice is **(3)**.

PART II

25. First make a chart for all integer values from –2 to 7:

x	$f(x)$
–2	$\sqrt{-2+2} = \sqrt{0} = 0$
–1	$\sqrt{-1+2} = \sqrt{1} = 1$
0	$\sqrt{0+2} = \sqrt{2} \approx 1.4$
1	$\sqrt{1+2} = \sqrt{3} \approx 1.7$
2	$\sqrt{2+2} = \sqrt{4} = 2$
3	$\sqrt{3+2} = \sqrt{5} \approx 2.2$
4	$\sqrt{4+2} = \sqrt{6} \approx 2.4$
5	$\sqrt{5+2} = \sqrt{7} \approx 2.6$
6	$\sqrt{6+2} = \sqrt{8} \approx 2.8$
7	$\sqrt{7+2} = \sqrt{9} = 3$

Plot the 10 points on the graph, and join the points with a curve.

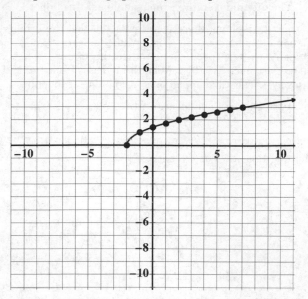

26. The graph of a nonlinear function is not a line. When the four ordered pairs are plotted, the graph looks like this:

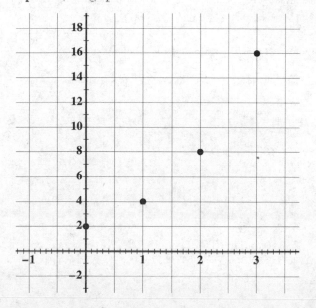

Since there is no line that passes through all four of these points, Caleb is correct. The ordered pairs are from a nonlinear function.

Even without graphing, there is a way to tell that the ordered pairs are from a nonlinear function. In a linear function every time x increases by 1, $f(x)$ increases by a specific amount. In this example when x increases from 0 to 1, $f(x)$ increases by 2. When x increases from 1 to 2, though, $f(x)$ increases by 4. This can happen only in a nonlinear function.

Caleb is correct. It is a nonlinear function.

27. Use the quadratic formula from the reference sheet, $x = \dfrac{-b \pm \sqrt{b^2 - 4ac}}{2a}$. For this question, $a = 1$, $b = 1$, and $c = -5$.

$$x = \frac{-1 \pm \sqrt{1^2 - 4 \cdot 1 \cdot (-5)}}{2 \cdot 1}$$

$$= \frac{-1 \pm \sqrt{1 + 20}}{2}$$

$$= \frac{-1 \pm \sqrt{21}}{2}$$

$$\frac{-1 + \sqrt{21}}{2} \approx 1.79 \approx 1.8$$

$$\frac{-1 - \sqrt{21}}{2} \approx -2.79 \approx -2.8$$

To the *nearest tenth*, $x = 1.8$ and $x = -2.8$.

28. The graph of $p(x + 2)$ is the graph of $p(x)$ shifted to the *left* by 2 units.

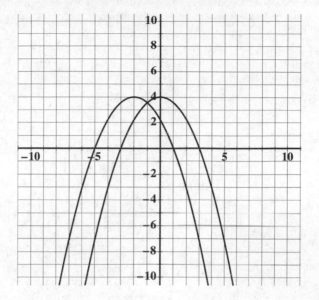

29. When the apple hits the ground, the height of the apple will be 0. To determine algebraically when this will happen, solve the equation $0 = -16t^2 + 256$. The answer must be positive since the number of seconds for the apple to fall cannot be negative:

$$0 = -16t^2 + 256$$
$$+16t^2 = +16t^2$$
$$\frac{16t^2}{16} = \frac{256}{16}$$
$$t^2 = 16$$
$$t = \sqrt{16} = 4$$

The apple takes 4 seconds to hit the ground.

30. Start with the distributive property. Then use the properties of algebra to get the constants on one side of the equals sign and the variables on the other side of the equals sign. To finish solving the equation, isolate the x variable:

$$6 - \frac{2}{3}(x+5) = 4x$$

$$6 - \frac{2}{3}x - \frac{2}{3} \cdot 5 = 4x$$

$$6 - \frac{2}{3}x - \frac{10}{3} = 4x$$

$$+\frac{2}{3}x = +\frac{2}{3}x$$

$$6 - \frac{10}{3} = 4x + \frac{2}{3}x$$

$$\frac{18}{3} - \frac{10}{3} = \frac{12}{3}x + \frac{2}{3}x$$

$$\frac{8}{3} = \frac{14}{3}x$$

$$\frac{3}{14} \cdot \frac{8}{3} = \frac{3}{14} \cdot \frac{14}{3}x$$

$$\frac{8}{14} = x$$

$$\frac{4}{7} = x$$

The exact value of x is $\frac{4}{7}$.

31. A rational number is one that can be written as a fraction that has an integer in both the numerator and the denominator. Since 16 is a perfect square, $\sqrt{16} = 4$ is a rational number. The product of 4 and $\frac{4}{7}$ is $4 \cdot \frac{4}{7} = \frac{16}{7}$. Since this fraction has an integer in both the numerator and the denominator, it is rational.

The product is rational.

32. Since the boundary between the two pieces is $x = 2$, make a chart with some x-values less than 2 and some x-values greater than 2. For $x = -1$, 0, and 1, use the first equation in the function. For $x = 2, 3, 4$, and 5, use the second equation in the function:

x	$f(x)$
-1	$-\dfrac{1}{2} \cdot (-1) = \dfrac{1}{2}$
0	$-\dfrac{1}{2} \cdot 0 = 0$
1	$-\dfrac{1}{2} \cdot 1 = -\dfrac{1}{2}$
2	2
3	3
4	4
5	5

This is what the 7 points look like on the graph:

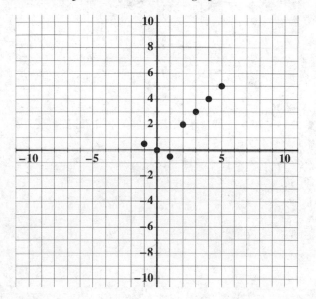

To plot the first piece of the graph, find the boundary point by plugging 2 into the first equation $-\frac{1}{2} \cdot 2 = -1$. This corresponds to the point $(2, -1)$.

Even though $(2, -1)$ is not an actual point on the graph, it is important to plot it on the graph as a small open circle. Make a ray that starts at this open circle and that passes through $\left(1, -\frac{1}{2}\right)$.

To plot the second piece of the the graph, make a ray that starts at $(2, 2)$ and that goes through $(3, 3)$.

The finished graph looks like this:

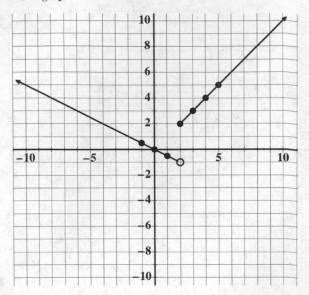

PART III

33. In an exponential equation of the form $y = a \cdot b^x$, a represents the starting amount and b represents the growth factor. In this question, the 20 represents the number of rabbits when the population was first counted. The 1.014 is the growth factor. The growth factor is also 1 plus the growth rate, so this population is increasing by 1.4% a year.

The average rate of change of a function $p(x)$ between $x = a$ and $x = b$ can be calculated with the formula $\dfrac{p(b) - p(a)}{b - a}$. In this question, $a = 50$ and $b = 100$. Calculate the average rate of change:

$$\frac{p(100) - p(50)}{100 - 50} = \frac{20(1.014)^{100} - 20(1.014)^{50}}{100 - 50}$$

$$= \frac{80.3 - 40.1}{100 - 50}$$

$$= \frac{40.2}{50}$$

$$= 0.804 \approx 0.8$$

To the *nearest tenth,* the average rate of change from day 50 to day 100 is 0.8.

34. For Garage A, the cost of parking for x hours (where x is greater than 2) is $A = 7 + 3(x - 2)$. For example, if you park for 6 hours in Garage A, the cost is $7 for the first 2 hours and then $3 for the additional <u>4</u> (not 6) hours. A common mistake is to write $A = 7 + 3x$ for the cost. However, that equation would make you pay an extra $3 for each of the first 2 hours.

For Garage B, the cost of parking for x hours is $B = 3.25x$.

The cost of parking at both garages will be the same when $7 + 3(x - 2) = 3.25x$.

Algebraically, this can be solved as follows:

$$7 + 3(x - 2) = 3.25x$$
$$7 + 3x - 6 = 3.25x$$
$$\underline{-3x = -3x}$$
$$7 - 6 = 3.25x - 3x$$
$$\frac{1}{0.25} = \frac{0.25x}{0.25}$$
$$4 = x$$

Both garages cost the same when you park for 4 hours.

35. Linear inequalities are graphed as a boundary line and all the points on one side of the line are shaded. To graph the inequality $2y + 3x \leq 14$, first graph the boundary line. Start by changing the inequality sign into an equals sign. Then rewrite the equation in slope-intercept form $y = mx + b$:

$$2y + 3x \leq 14$$

$$2y + 3x = 14$$

$$-3x = -3x$$

$$\frac{2y}{2} = \frac{-3x + 14}{2}$$

$$y = -\frac{3}{2}x + 7$$

Since the inequality has a $\leq$ instead of a $<$, graph a solid line.

To determine which side of the line to shade, test if the point $(0, 0)$ makes the inequality true. Substitute $(0, 0)$ into the original inequality $2y + 3x \leq 14$:

$$2 \cdot 0 + 3 \cdot 0 \leq 14$$

$$0 + 0 \leq 14$$

$$0 \leq 14$$

Since it is true that 0 is less than or equal to 14, shade the side of the boundary line that contains $(0, 0)$.

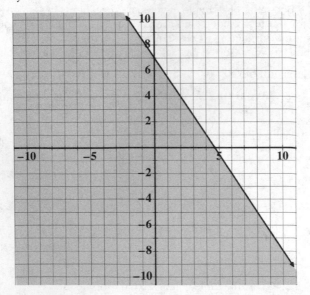

To find the boundary line, change the inequality $4x - y < 2$ into an equation by substituting an equals sign for the inequality sign. Then write the equation in slope-intercept form $y = mx + b$:

$$4x - y < 2$$

$$4x - y = 2$$

$$-4x = -4x$$

$$\frac{-y}{-1} = \frac{-4x + 2}{-1}$$

$$y - 4x - 2$$

Since the inequality has a < instead of a ≤, graph a dotted line.

Test to see if $(0, 0)$ is a solution to the second inequality:

$$4 \cdot 0 - 0 < 2$$

$$0 - 0 < 2$$

$$0 < 2$$

Since this is true, shade the side of the line that contains $(0, 0)$.

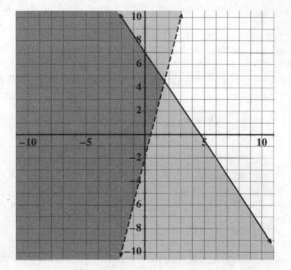

Inequalities can also be graphed using the graphing calculator.

For the TI-84:

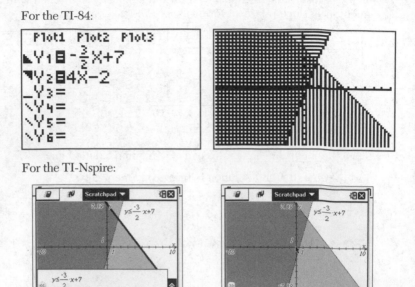

For the TI-Nspire:

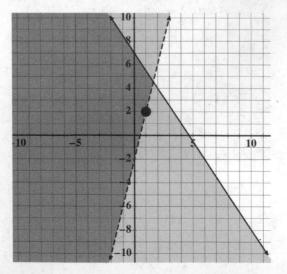

A point is in the solution set if it is in the double-shaded area. The point $(1, 2)$ is on the dotted line, so it is not part of the double-shaded area and, therefore, is not part of the solution set. If it was in the double-shaded area or even on the solid boundary line, it would be part of the solution set.

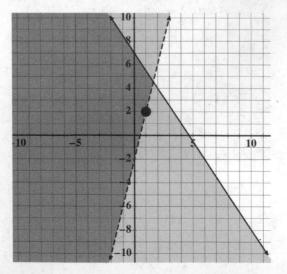

36. Enter the data into the graphing calculator, and use the linear regression function.

For the TI-84:

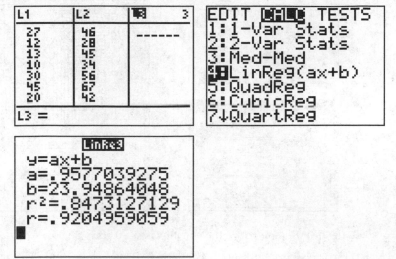

If your TI-84 calculator is not showing the value of r, which is the correlation coefficient, you have to turn on the diagnostics by pressing [2nd] [0] and scrolling down to DiagnosticOn.

For the TI-Nspire:

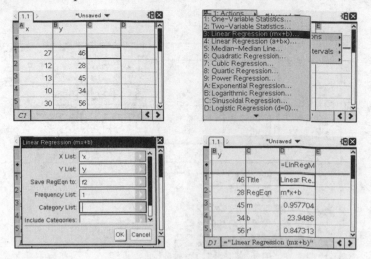

To the *nearest hundredth*, the equation is $y = 0.96x + 23.95$.

The correlation coefficient is 0.92, which is very close to +1. This means that there is a strong positive correlation. If the correlation coefficient was very close to –1, there would be a strong negative correlation. If the correlation coefficient was close to 0, there would be a weak correlation.

PART IV

37. If d is the number of dimes and q is the number of quarters, the following system of equations can model this situation:

$$0.10d + 0.25q = 17.55$$
$$d + q = 90$$

To solve this system of equations, start by isolating the d in the second equation. Then use the substitution method so the only remaining variable is q, which shows the number of quarters Dylan has in his bank:

$$d = 90 - q$$
$$0.10d + 0.25q = 17.55$$
$$0.10(90 - q) + 0.25q = 17.55$$
$$9 - 0.10q + 0.25q = 17.55$$
$$-9 = -9$$
$$-0.10q + 0.25q = 17.55 - 9$$
$$\frac{0.15q}{0.15} = \frac{8.55}{0.15}$$
$$q = 57$$

Dylan has 57 quarters in his bank.

If each of Dylan's dimes is replaced with a quarter, he would have 90 quarters. They would be worth:

$$90 \times \$0.25 = \$22.50$$

The game costs $20.98 plus 8% of $20.98 for sales tax:

$$\$20.98 + (0.08 \times \$20.98) = \$22.66$$

Including sales tax, the game costs $22.66, which is more than $22.50.

No, Dylan would not be able to buy the game if all his dimes are replaced with quarters.

Topic	Question Numbers	Number of Points	Your Points	Your Percentage
1. Polynomials	3, 12, 19,	2 + 2 + 2 = 6		
2. Properties of Algebra	30	2		
3. Functions	2, 8, 11, 20	2 + 2 + 2 + 2 = 8		
4. Creating and Interpreting Equations	6, 14, 17, 21, 23	2 + 2 + 2 + 2 + 2 = 10		
5. Inequalities	1, 35	2 + 4 = 6		
6. Sequences and Series	7, 24	2 + 2 = 4		
7. Systems of Equations	34, 37	4 + 6 = 10		
8. Quadratic Equations and Factoring	4, 10, 22, 27, 29	2 + 2 + 2 + 2 + 2 = 10		
9. Regression	36	4		
10. Exponential Equations	26, 33	2 + 4 = 6		
11. Graphing	13, 16, 18, 25, 28, 32	2 + 2 + 2 + 2 + 2 + 2 = 12		
12. Statistics	5, 9	2 + 2 = 4		
13. Number Properties	31	2		
14. Unit Conversions	15	2		

HOW TO CONVERT YOUR RAW SCORE TO YOUR ALGEBRA I REGENTS EXAMINATION SCORE

The accompanying conversion chart must be used to determine your final score on the June 2018 Regents Examination in Algebra I. To find your final exam score, locate in the column labeled "Raw Score" the total number of points you scored out of a possible 86 points. Since partial credit is allowed in Parts II, III, and IV of the test, you may need to approximate the credit you would receive for a solution that is not completely correct. Then locate in the adjacent column to the right the scale score that corresponds to your raw score. The scale score is your final Algebra I Regents Examination score.

Regents Examination in Algebra I—June 2018
Chart for Converting Total Test Raw Scores to Final
Examination Scores (Scaled Scores)

Raw Score	Scale Score	Performance Level	Raw Score	Scale Score	Performance Level	Raw Score	Scale Score	Performance Level
86	100	5	57	81	4	28	67	3
85	99	5	56	81	4	27	66	3
84	98	5	55	81	4	26	65	4
83	97	5	54	81	4	25	63	2
82	96	5	53	80	4	24	62	2
81	94	5	52	80	4	23	60	2
80	93	5	51	80	4	22	59	2
79	93	5	50	80	4	21	57	2
78	92	5	49	79	3	20	56	2
77	91	5	48	79	3	19	55	2
76	90	5	47	79	3	18	52	1
75	89	5	46	78	3	17	51	1
74	89	5	45	78	3	16	49	1
73	88	5	44	78	3	15	47	1
72	87	5	43	77	3	14	44	1
71	87	5	42	77	3	13	42	1
70	86	5	41	76	3	12	40	1
69	86	5	40	76	3	11	37	1
68	86	5	39	75	3	10	35	1
67	85	5	38	75	3	9	32	1
66	84	4	37	74	3	8	29	1
65	84	4	36	74	3	7	26	1
64	84	4	35	73	3	6	23	1
63	83	4	34	72	3	5	20	1
62	83	4	33	71	3	4	16	1
61	83	4	32	71	3	3	13	1
60	82	4	31	70	3	2	9	1
59	82	4	30	69	3	1	5	1
58	82	4	29	68	3	0	0	1

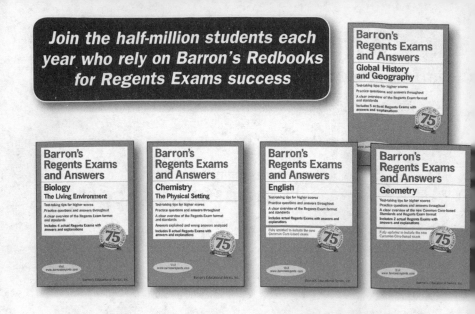